U0193193

中华二十四节气菜

川菜卷

CHINESE 24 SOLAR TERM DISHES
(SICHUAN CUISINE)

中英文对照版

主编◇杜莉 陈祖明

摄影◇程蓉伟

四川科学技术出版社

图书在版编目（CIP）数据

中华二十四节气菜（川菜卷）= Chinese 24 Solar Term Dishes (Sichuan Cuisine) : 汉英对照 / 杜莉，陈祖明主编；程蓉伟图片摄影. —成都：四川科学技术出版社，2020.9

ISBN 978-7-5364-9949-2

Ⅰ.①中… Ⅱ.①杜… ②陈… ③程… Ⅲ.①食物养生—菜谱—汉、英②川菜—菜谱—汉、英 Ⅳ.①R247.1 ②TS972.161

中国版本图书馆CIP数据核字(2020)第179828号

中英文对照版

中华二十四节气菜（川菜卷）

CHINESE 24 SOLAR TERM DISHES (SICHUAN CUISINE)

主编：杜莉　陈祖明　　摄影：程蓉伟

出 品 人	程佳月
责任编辑	程蓉伟
出版发行	四川科学技术出版社
装帧设计	程蓉伟
封面设计	程蓉伟
责任印制	欧晓春
制　　作	成都华桐美术设计有限公司
印　　刷	成都市金雅迪彩色印刷有限公司
成品尺寸	285mm×210mm
印　　张	21.5
字　　数	300千
版　　次	2020年9月第1版
印　　次	2020年9月第1次印刷
书　　号	ISBN 978-7-5364-9949-2
定　　价	258.00元

二十四节气歌

春雨惊春清谷天，夏满芒夏暑相连。
秋处露秋寒霜降，冬雪雪冬小大寒。
每月两节不变更，最多相差一两天。
上半年来六廿一，下半年是八廿三。

《中华二十四节气菜》（川菜卷）编纂委员会

顾　问

吴　旭　卢　一　王荣木　陈莉萍

主　任

赵文峤　许志勋　周月霞　陈云川

副主任

孙　智　余贵冰　邓爱华　陈锦阳　杜　莉　李　想　陈祖明　张　媛

委　员

（以姓氏笔画为序）

王胜鹏　冉　伟　冯明会　朱　健　乔　兴　刘军丽　李　力　李平南　吴先锋
张　茜　张　胜　陈　龙　陈　筱　陈丽兰　孙晋康　罗　文　罗　刚　郑　伟
钟志惠　禹　伟　聂瑞刚　徐向波　程蓉伟　詹　珂　詹淞丞

主　编

杜　莉　陈祖明

副主编

张　茜　王胜鹏　罗　文　乔　兴　冉　伟

创意菜肴设计制作

冉　伟　乔　兴　陈祖明　詹淞丞　陈丽兰

创意面点小吃设计制作

罗　文　冯明会　徐向波

中文撰稿及审校

杜　莉　陈祖明　张　茜　詹　珂　王胜鹏　刘军丽　郑　伟
罗　文　冉　伟　乔　兴　詹淞丞　冯明会　徐向波　贺　树

英文翻译及审校

张　媛　尹　川　李潇潇　彭施龙　王　畅　张　琪

图片摄影及菜点造型

程蓉伟　陈　筱

编纂单位

资阳市人民政府　四川旅游学院

Compilation Committee of *Chinese 24 Solar Term Dishes (Sichuan Cuisine)*

Counselors

Wu Xu, Lu Yi, Wang Rongmu, Chen Liping

Directors

Zhao Wenqiao, Xu Zhixun, Zhou Yuexia, Chen Yunchuan

Deputy Directors

Sun Zhi, Yu Guibing, Deng Aihua, Chen Jinyang, Du Li, Li Xiang, Chen Zuming, Zhang Yuan

Committee Members

Wang Shengpeng, Ran Wei, Feng Minghui, Zhu Jian, Qiao Xing, Liu Junli, Li Li, Li Pingnan, Wu Xianfeng, Zhang Qian, Zhang Sheng, Chen Long, Chen Xiao, Chen Lilan, Sun Jinkang, Luo Wen, Luo Gang, Zheng Wei, Zhong Zhihui, Yu Wei, Nie Ruigang, Xu Xiangbo, Cheng Rongwei, Zhan Ke, Zhan Songcheng

Editors-in-Chief

Du Li, Chen Zuming

Deputy Editors- in- Chief

Zhang Qian, Wang Shengpeng, Luo Wen, Qiao Xing, Ran Wei

Design and Production of Creative Dishes

Ran Wei, Qiao Xing, Chen Zuming, Zhan Songcheng, Chen Lilan

Design and Production of Creative Pastry Snacks

Luo Wen, Feng Minghui, Xu Xiangbo

Chinese Compiled and Reviewed by

Du Li, Chen Zuming, Zhang Qian, Zhan Ke, Wang Shengpeng, Liu Junli, Zheng Wei, Luo Wen, Ran Wei, Qiao Xing, Zhan Songcheng, Feng Minghui, Xu Xiangbo, He Shu

English Translated and Reviewed by

Zhang Yuan, Yin Chuan, Li Xiaoxiao, Peng Shilong, Wang Chang, Zhang Qi

Photography and Dish Modeling

Cheng Rongwei, Chen Xiao

Compilation Units

Ziyang Municipal People`s Government, Sichuan Tourism University

序

文化之根基，在于我们脚下的土地；文化之成就，在于大众的探索、创造、总结和传承。巴蜀文化的繁荣，在于深耕于巴蜀大地和对外来文化的兼收并蓄及不断扬弃。

二十四节气是中华文明中的一朵灿烂之花，集天文、地理和农业之大成，十分科学，是我国农业生产必须遵循的时间节律。二十四节气的完整记载始见于西汉初年淮南王刘安的《淮南子》，而官方正式颁布二十四节气是在汉武帝太初元年（公元前104年）的《太初历》，该历法是由四川阆中人落下闳组织制定。二十四节气确立至今历经两千多年，一直指引着我国的农业生产和人们的日常生活，凝聚着中华民族的科学智慧与文化精髓，是中华民族对人类文明的伟大贡献，并于2016年被列入联合国教科文组织人类非物质文化遗产代表作名录。

二十四节气中的每个节气都表示着时候、气候、物候等“三候”的不同变化，智慧的中国人民按照天人合一、适应自然的规律，形成了与二十四节气相伴的一系列民俗习惯，包括农耕、着衣、饮食、养生和祭祀活动。就饮食而言，因为不同节气的变化，自然生长的食材也出现了相应的周期性变化，而中华民族在长期的饮食生活中遵循天人相应、辨证施食等饮食观念，便不断创造出一些与不同节气相应的独特饮食习俗和菜点。随着“二十四节气”被列入联合国教科文组织人类非物质文化遗产代表作名录，我国各地积极开展节气菜的挖掘整理、传承创新，中华二十四节气菜体系正在形成。

庚子年初，新冠肺炎疫情暴发，给我国的社会经济带来了严重冲击，世界格局也发生了深刻变化，餐饮业更是损失巨大。时逢国家层面积极推动成渝地区双城经济圈建设战略，给四川和重庆化危为机及高质量发展带来了重大机遇。在这样的历史背景下，世界中餐业联合会、资阳市人民政府、四川旅游学院等联合举办“中华节气菜大会”，对于提振餐饮业信心，促进社会经济的恢复与发展，弘扬优秀传统文化均有着十分重要的意义。期间，由四川旅游学院杜莉教授、陈祖明教授率领项目团队主编的《中华二十四节气菜（川菜卷）》一书将与社会各界见面。该书不仅系统梳理了中华二十四节气的文化内涵，还深入挖掘了四川省特别是资阳市二十四节气的饮食习俗、文化传统，并以此为依托，以资阳及四川各地各季的代表性原料为主要食材，研发、创制了具有浓郁地方特色的川味二十四节气菜。

可以说，《中华二十四节气菜（川菜卷）》的编撰、出版是惠泽大众的有益之事，有助于提升资阳美食及川菜的文化软实力和国际形象，同时对更好地传承中华饮食文明也有极好的示范作用。希望不久的将来，通过全国餐饮人的辛勤耕耘，中华二十四节气菜的百花园必将花开满园，香飘世界。

是为序。

庚子年九月于青城读味楼

PREFACE

The foundation of culture lies in the land under our feet; the achievement of culture lies in the exploration, creation, summary and inheritance of people. The prosperity of Bashu culture lies in its deep cultivation in Bashu land and its eclecticism of foreign cultures, and its constant sublation.

The 24 Solar Terms are a brilliant flower in Chinese civilization. They are the great achievements of astronomy, geography and agriculture. They are very scientific. They are the time rhythm that our agricultural production must follow. The complete records of the 24 Solar Terms can be found in *Huai Nan Tzu* written by Liu An, the king of Huainan in the early Western Han Dynasty.

However, the official promulgation of the 24 Slar Terms was in the *Taichu Calendar* in the first year of Taichu (104 BC) of Emperor Wu of Han Dynasty. The calendar was formulated by Lao Xiahong of Langzhong, Sichuan Province. It has been more than 2,000 years since the establishment of the 24 Solar Terms, which have been guiding China's agricultural production and people's daily life, condensing the scientific wisdom and cultural essence of the Chinese nation. It is a great contribution of the Chinese nation to human civilization. It has been listed in the UNESCO Representative List of the Intangible Cultural Heritage of Humanity since 2016.

Each solar term in the 24 Solar Terms represents the different changes of time, climate and phenology. According to the unity of man and nature, adapting to the laws of nature, the wise Chinese people have formed a series of folk customs associated with the 24 Solar Terms, including farming, clothing, drinking, health preservation and God worship.

As far as diet is concerned, due to the changes of different solar terms, the natural growth of food materials also has corresponding periodic changes. In the long-term diet life, the Chinese people follow the dietary concepts of "correspondence between man and nature" and "dialectical feeding", so they have constantly created some unique eating customs and dishes corresponding to different solar terms. With the "24 Solar Terms" listed in the World Intangible Cultural Heritage List, China's 24 Solar Terms system is to be developing.

The outbreak of the COVID-19 at the beginning of the Gengzi Rat Year has brought serious impact to our economy and society, and the world pattern has undergone profound changes. At the same time, the national level has greatly promoted the construction strategy of double city economic circle in Chengdu-Chongqing region, which has brought great opportunities to Sichuan and Chongqing of turning crisis into opportunity and high-quality development.

Under this historical background, the World Federation of Chinese Food Industry, Ziyang Municipal People's Government, Sichuan Tourism University, etc. jointly hold the Chinese Solar Term Dish Conference, which is of great significance to boost the confidence of the catering industry, promote the economic and social recovery and development, and carry forward the excellent traditional culture.

During this period, the book *Chinese 24 Solar Term Dishes (Sichuan Cuisine)* will meet readers, which is edited by the project team led by Professor Du Li and Professor Chen Zuming of Sichuan Tourism University. The book not only systematically combs the cultural connotation of the 24 Solar Terms in China, but also deeply excavates the food customs and cultural traditions of Sichuan Province, especially Ziyang City. Based on this, the book develops and creates 24 Solar Terms dishes in Sichuan flavor, with strong local characteristics based on the representative raw materials of Ziyang and other seasons in Sichuan Province.

It can be said that the compilation and publication of *Chinese 24 Solar Term Dishes (Sichuan Cuisine)* is beneficial to the public, which helps to enhance the cultural spread and international image of Ziyang cuisine and Sichuan cuisine, and is an excellent demonstration in inheriting Chinese food civilization. I hope that in the near future, through the hard work of the national catering people, the Chinese 24 Solar Terms dishes will bloom all over the world.

This serves the preface.

Lu Yi

Duwei Building, Mount Qingcheng, September of the Year of Gengzi

前言

作为一个农业大国，中国自古以来就非常重视农业生产。为了提高农业产量，能够准确判断天时就显得至关重要。经过长时间的观察、实践，我国先民逐渐总结出了天气变化的规律，并制定出二十四节气用于指导农业生产。二十四节气作为我国传统的天文与人文合一的历法现象，延续至今已数千年，是中华民族珍贵的文化与科学遗产。2016年11月，二十四节气被列入联合国教科文组织人类非物质文化遗产代表作名录。

所谓二十四节气，指的是干支历中表示自然节律变化以及确立“十二月建”的特定节令，始于立春，终于大寒。二十四节气是上古先民顺应农时，通过观察天体运行，认知一岁（年）中时候（时令）、气候、物候变化规律所形成的知识体系，每个节气都表示着时候、气候、物候等“三候”的不同变化。二十四节气更表达了人与自然宇宙之间独特的时间观念，蕴含着中华民族悠久的文化内涵和历史积淀，不仅在农业生产上起着指导作用，还影响着中国人的衣食住行及文化观念，形成了中华民族特有的节令文化和独特的饮食养生习俗，如立春时节“贴春联”“咬春”，立秋时节“贴秋膘”，等等。在四川，节气饮食习俗也较为丰富多样，如立秋时民间有“尝新”习俗。立秋之后，四川许多地方水稻收获，人们将刚收获的水稻加工成新米、煮成新米饭后品尝。在“尝新”之前，旧俗还要先盛新米饭敬献祖宗、谷神和土地。又如四川一些地方有“冬至抄手夏至面”的饮食习俗，在川西平原，冬至时也有吃羊肉汤的习惯。中国人在饮食上讲究天人相应、辨证施食与五味调和，强调“时序为美”“不时不

食”，即遵循自然规律，根据不同季节时令，选择相应的食材制作和食用菜点，“二十四节气菜”由此而生。

2020年10月，首届“中华节气菜大会”将在资阳市召开。它是继2019年“第二届世界川菜大会”之后资阳美食产业发展过程中的又一次盛会，更是新冠肺炎疫情重创餐饮业后提振餐饮业信心、推动餐饮业恢复和高质量发展的促进大会。资阳地处天府之国的中西部、沱江西岸，是成渝地区双城经济圈的四川省区域性中心城市。资阳境内气候湿润，地形多样，河流纵横，自古便是四川盆地农业发达的重要区域之一，出产丰富的优质食材。资阳依托丰富的自然资源和悠久的历史文化，在政府引导、协会助推、市场主导、企业运作、立足民间等原则的指导下，创造了特色鲜明、异彩纷呈的美食文化，也成为了四川二十四节气饮食的典型代表。2019年，四川旅游学院与资阳市商务局合作，编撰出版了《资阳美食文化》一书，全面、系统地梳理和总结了资阳美食文化的整体情况，取得了良好的效果。2020年，双方以首届“中华节气菜大会”为契机，再次携手合作，编撰出版《中华二十四节气菜（川菜卷）》一书。

本书由四川旅游学院川菜发展研究中心、资阳市商务局共同负责，组建了包括饮食文化、食疗养生、烹饪技术、美食创意、英语翻译等方面的专家学者、烹饪大师及名师、艺术设计者、翻译人员在内的项目团队共同开展工作。团队通过文献搜集、实地调研等方式，搜集、整理了从古至今有关二十四节气饮食的资料，特别注重深入挖掘资阳以及整个四川的

前言

二十四节气饮食养生民俗等资料，并以此为依托，设计全书框架大纲，经反复征求多方意见与建议进行完善。此后，团队在编撰全书初稿的同时，引入中华二十四节气的饮食习俗、养生理论、味觉艺术，以及安全、健康、特色等现代消费理念，结合资阳市及整个四川二十四节气饮食养生习俗和文化传统，以当季、当地代表性原料为主要食材，在传承的基础上改良、研发出川味浓郁的二十四节气菜（资阳节气菜48道、其他川菜品种24道），并进行图片拍摄、文稿撰写，充实到全书之中，再提交双方领导、专家征求意见和建议，最后形成定稿，进行英文翻译。

《中华二十四节气菜（川菜卷）》一书由杜莉设计框架大纲并负责总体统筹。全书分为上篇和下篇，上篇从二十四节气的起源与发展以及川菜食材、川菜养生、四川食俗、四川茶酒等五个方面进行概括性阐述；下篇则分别以每个节气为对象，分两个部分阐述：第一部分阐述每个节气的诗文、物候、食材生产、饮食养生和饮食习俗等，第二部分阐述每个节气所研发的创意菜点。其中，上篇和下篇中的第一部分文稿主要由王胜鹏、张茜、郑伟、刘军丽、詹珂等执笔，二十四节气菜点的创意研发和下篇中的第二部分文稿主要由陈祖明、罗文、冉伟、冯明会、乔兴、徐向波、詹淞丞、陈丽兰等完成。最后由杜莉统稿、审核中文稿，张茜、贺树校正，由张媛、尹川、李潇潇、彭施龙、王畅、张琪等翻译成英文并审校。由程蓉伟对创意菜点进行装盘设计并摄影，陈筱参与装盘设计并负责资料汇总与沟通协调等工作。全书内容丰富，图文并茂，中英文对照，集学术性、实用性、趣味性于一体，不仅能够大力传承及弘扬中国优秀传统文化，让世界更好地了解中国，推动中国文化及饮食文化的国际传播，而且还可促进季节性、地方性特色菜点的研制和有序转换，也有利于推动资阳市乃至四川地方特色美食和旅游业更好地发展。

本书是四川旅游学院、资阳市政府相关部门和四川科学技术出版社共同努力的结晶。在研究和编撰过程中，不仅得到了四川旅游学院领导、资阳市政府领导的高度重视和指导，也得到了资阳市商务局领导和四川科学技术出版社领导的大力支持，为本书的编撰、出版奠定了坚实基础。在本书顺利出版之际，我们对给予项目团队关心、指导、支持和帮助的各位领导、专家学者等表示衷心感谢！同时，也对所有参与此项工作、付出辛勤劳动与宝贵节假日的朋友表示衷心感谢！

中华二十四节气的文化内涵丰厚，但是，由于开展本书编撰、出版工作的时间十分紧迫，加之能力所限，书中难免有不足甚至错漏之处，敬请广大读者不吝赐教，以便今后修订和完善。

编　者

2020年9月于成都

FORWARD

As a big agricultural country, China has attached great importance to agricultural production since ancient times. In order to improve agricultural production, it is very important to judge the weather accurately. After a long period of observation and practice, our ancestors gradually summed up the law of weather change, and formulated 24 Solar Terms to guide agricultural production. The 24 Solar Terms, as a traditional astronomical and humanistic calendar phenomenon, has lasted for thousands of years and is a precious cultural and scientific heritage of China. In November 2016, the 24 Solar Terms has been listed in the UNESCO Representative List of the Intangible Cultural Heritage of Humanity.

The 24 Solar Terms are the special terms in the lunar calendar which show the change of nature and fix the words defining the months. The 24 Solar Terms start from the Beginning of Spring and end in Greater Cold. The 24 Solar Terms are the knowledge hierarchy. In order to create the knowledge hierarchy, the ancient people follow farming season and recognize time, climate, phenology by observing celestial motion. Every term means the different change of time, climate and phenology. The 24 Solar Terms show the time concept between human and universe, contain the long time cultural connotation and historical accumulation of Chinese nation. The 24 Solar Terms not only guide the agricultural production, but also affect the daily life and cultural concept of Chinese people. It has created the special terms culture and dietary regimen of Chinese people, like pasting spring poem and biting spring in the Beginning of Spring, sticking autumn fat in the Beginning of Autumn, etc. In Sichuan, the eating customs of terms are colourful. For example, there is custom of tasting a fresh delicacy in the Beginning of Autumn. After the Beginning of Autumn, grains have been harvested in lots of parts in Sichuan. People will make fresh rice and cook fresh rice with the newly harvested grains. After tasting a fresh delicacy, people would offer cooked rice to ancestors, god of grain and the land in ancient time. In some parts of Sichuan, there is an eating custom of eating Chao Shou in the Winter Solstice and noodle in the Summer Solstice. In Western Sichuan Plain, there is a custom of eating mutton soup in the Winter

Solstice. On eating, Chinese people devote particular care to correspondence between man and universe, dialectical feeding, harmony of five flavors; emphasize on eating in proper time, following the nature order, choosing proper food ingredients to make dishes for eating according to different seasons and terms. Here comes the 24 Solar Terms dishes.

In 0ctober, 2020, Ziyang City will host the first Chinese 24 Solar Term Dishes Conference. It is another great event after The 2nd World Sichuan Cuisine Conference. It is a conference which can promote the confidence of catering industry, recover and develop the catering industry from the impact of COVID-19. Ziyang City is located in the midwest of the land of abundanc and the west bank of Tuojiang River. It is the regional center of Chengdu-Chongqing economic circle. Ziyang City has the humid climate, varied terrain, crisscross rivers. It is an important developed agricultural area which can produce abundant good food ingredients since ancient time. Relying on abundant natural resources and long history and culture, Ziyang City has created colorful and charming cuisine culture under the principle of government leading, association promoting, market orienting, enterprise operating and normal people supporting. Ziyang City has become the representative of Sichuan's 24 Solar Terms diet.

In 2019, Sichuan Tourism University has cooperated with Ziyang Municipal Bureau of Commerce to publish the book *Ziyang Food Culture*. The book has teased and summarized the food culture of Ziyang and get a good feedback. In 2020, the two sides will cooperate again to publish the *Chinese 24 Solar Term Dishes (Sichuan Cuisine)* for the first Chinese 24 Solar Term Dishes Conference.

This book is jointly responsible by Sichuan Cuisine Development Research Center of Sichuan Tourism University and Ziyang Municipal Bureau of Commerce. It has established a project team including experts and scholars in food culture, diet therapy and health preservation, cooking technology, food creativity and English translation. The team also include culinary masters, art designers and translators. Through literature collection and field research, the team collected and sorted out the materials about the 24 Solar Terms diet from ancient times to the present, especially focused on the excavation of the 24 Solar Terms diet and health preservation folk custom of Ziyang and Sichuan Province.

Based on this, the framework of the book was designed and perfected after repeatedly soliciting opinions and suggestions. After that, while compiling the first draft of the book, the team introduced modern consumption concepts such as Chinese 24 Solar Terms diet custom, health preservation theory, taste art and its safety, health and characteristics. Combining 24 Solar Terms diet regimen customs and cultural traditions in Ziyang and Sichuan Province, taking the seasonal and local representative raw materials as the main ingredients, we improved and developed the 24 Solar Terms dishes (48 Ziyang solar terms dishes and other 24 Sichuan dishes). We wrote down and enriched them into the book, and then submitted to leaders and experts of both sides for comments and suggestions. Finally, the final draft was formed and translated into English.

The book *Chinese 24 Solar Term Dishes (Sichuan Cuisine),* is designed by Du Li and she is responsible for overall planning, which is divided into Part One and Part Two. In Part One, this book discusses five aspects: the origin and development of the 24 Solar Terms, Sichuan food materials, Sichuan food regimen, Sichuan food customs, and Sichuan tea and liquor.

Part Two goes deeper into each solar term, and is divided into two parts: the first part describes the poetry, phenology, food material production, diet regimen and diet customs of each solar term, and the second part describes the creative dishes developed by each solar term. Among them, the first part are mainly written by Wang Shengpeng, Zhang Qian, Zheng Wei, Liu Junli, Zhan Ke, etc. The creative research and development of 24 Solar Terms dishes and the second part of Part Two are mainly completed by Chen Zuming, Luo Wen, Ran Wei, Feng Minghui, Qiao Xing, Xu Xiangbo, Zhan Songcheng, Chen Lilan, etc. Finally, the Chinese version was unified and reviewed by Du Li, corrected by Zhang Qian and He Shu, and translated into English by Zhang Yuan, Yin Chuan, Li Xiaoxiao, Peng Shilong, Wang Chang and Zhang Qi. Cheng Rongwei designed and photographed creative dishes, and

Chen Xiao was responsible for information collection, communication and coordination. The book is rich in content, illustrated and written in both Chinese and English. It is academic, practical and interesting. It can not only inherit and carry forward traditional Chinese culture, but also let the world understand China better, as well as promote the international dissemination of Chinese culture and our catering culture. This book also promotes the development and orderly transformation of seasonal and local specialty dishes, and promotes special food and tourism developed in Ziyang and even in Sichuan.

This book is the fruit of the joint efforts of Sichuan Tourism University, relevant departments of Ziyang Municipal People's Government and Sichuan Publishing House of Science and Technology. In the process of research and compilation, the leaders of Sichuan Tourism University and Ziyang Municipal People's Government attached great importance to this book. It also got strong support from the leaders of Ziyang Municipal Bureau of Commerce and Sichuan Publishing House of Science and Technology, which laid a solid foundation for the compilation and publication of this book. On the coming publication of this book, I would like to express my heartfelt thanks to all the leaders, experts and scholars who care, guide, support and help the project team! At the same time, I would like to express my heartfelt thanks to all the friends who have participated in this work, and paid hard work and holidays on it!

The cultural connotation of the 24 Solar Terms in China is rich. However, due to the urgent time for compiling and publishing this book and the limitation of ability, there are inevitably deficiencies and even mistakes in the book. We sincerely hope that readers will give us your advice so as to revise and improve it in the future.

Compiler

September, 2020, Chengdu

目录

上篇 二十四节气与川菜概说

下篇 二十四节气与川菜创意菜点

CONTENTS

PART❶ 24 SOLAR TERMS AND SICHUAN CUISINE

PART❷ 24 SOLAR TERMS AND CREATIVE SICHUAN DISHES

SPRING

SUMMER

AUTUMN

WINTER

春之德风，风不信，其华不盛，华不盛，则果实不生。

夏之德暑，暑不信，其土不肥，土不肥，则长遂不精。

秋之德雨，雨不信，其谷不坚，谷不坚，则五种不成。

冬之德寒，寒不信，其地不刚，地不刚，则冻闭不开。

雨
風
順
調

PART 1 上篇 二十四节气与川菜概说

24 SOLAR TERMS AND SICHUAN CUISINE

二十四节气的起源与发展

ORIGIN AND DEVELOPMENT OF THE 24 SOLAR TERMS

二十四节气是中国传统的天文与人文合一的历法现象和独特的时间法则，至今已延续了数千年。它的出现与中国古代农业生产密切相关，是中华民族农耕文明的结晶。古代中国十分重视农业生产，然而，由于早期生产力水平非常低下，使得农业生产极大地受制于自然环境的影响。为了提高农业产量，能够准确判断天时便显得尤为重要，由此迫使中国的先民们很早就开始观察和研究天象气候的变化。经过长时间的观察、实践，中国先民逐渐总结出了天气变化的规律，并制定出二十四节气来指导农业生产。然而，二十四节气的出现不是一蹴而就的，它的划分也不是一次性完成的，而是随着人们对于气候感知的加深以及观测技术的进步而不断完善的。总体上说，二十四节气的历史沿革经历了萌芽、发展、定型完善、传承弘扬四个时期。

The 24 Solar Terms, which have been followed for several thousand years, are the calendar with the unification of Chinese traditional astronomy and humanity, and the particular law of time. Its appearance is closely connected with ancient Chinese agriculture, and it is crystallization of Chinese farming civilization. Ancient China has paid great attention to agricultural production. However, the agricultural production has been deeply influenced by natural environment because of the limitation of early productivity. In order to raise production, it is very important to estimate weather and climate precisely. Hence, ancestors have started to watch and analyze the changes of weather long time ago. On the base of long time practice, our ancestors have summarized the law of weather changing and have made the 24 Solar Terms to guide agricultural production. However, the 24 Solar Terms are not fixed overnight and are not finished in one time. They have been improved all the time with the development of technology and people’s perception of climate. Generally speaking, the history of the 24 Solar Terms has experienced the following 4 stages: budding stage, developing stage, finalizing stage and carrying forward stage.

一、萌芽时期：“分至四时” 的出现

二十四节气中，最早出现的是春分、夏至、秋分、冬至。据专家考证，中国古人很早就发明了“圭表测日”的方法。通过持续观测一年之中日影的变化，古人们最终发现了日影最长的“冬至日”和日影最短的“夏至日”这两个极点，并进而掌握了一年日影变化的周期性。根据《周髀算经》的记载可知，在后来的中国古代时间体系形成过程中，“圭表测日”同样发挥了极为重要的作用，它是古代划分二十四节气时间刻度最主要的方法。

Budding Stage: Emergence of "Liang Fen Liang Zhi"

In the 24 Solar Terms, the Spring Equinox, the Summer Solstice, the Autumn Equinox and the the Winter Solstice first appeared. According to textual research, ancient Chinese have invented the way of surveying sun shadow with gnomon long time ago. By observing the change of sun shadow persistently in a year, ancient people have found the day of the Winter Solstice when the sun shadow is the longest and the day of the Summer Solstice when the sun shadow is the shortest, and then have found the periodicity of the change in a year. According to the record in the book *Zhou Bi Suan Jing*, gnomon has played a very important role in the forming of ancient Chinese time system, and it is the main way of dividing time in the 24 Solar Terms

有关春、夏、秋、冬四时之分的文字记载，最早始见于西周时期的《尚书·尧典》。该书言："日中，星鸟，以殷仲春"；"日永，星火，以正仲夏"；"宵中，星虚，以殷仲秋"；"日短，星昴，以正仲冬"；"期三百有六旬有六日，……成岁"。"日"意为白天，"宵"意为夜晚，"中"意为平分。"日中"与"宵中"指的是昼夜相等的两天，即春分和秋分；"日永"和"日短"分别是白天最长和最短的两天，即夏至和冬至。整段话的意思是说，以春分昼夜平分之时和鸟星见于南方正中之时，作为考定仲春的依据；以夏至白昼最长之时和火星见于南方正中之时，作为考定仲夏的依据；以秋分昼夜平分之时和虚星见于南方正中之时，作为考定仲秋的依据；以冬至白昼最短之时和昴星见于南方正中之时，作为考定仲冬的依据，并且期望以三百六十六日为一个周期而形成一年。由此可见，在西周时期，我国先民们就已经通过对天气变化的长期系统观察，产生了"分至四时"的认知，即"两分两至"（春分、秋分与夏至、冬至），并且把冷暖交替的366天定为一年。

The earliest written record of four seasons is in the book *Shang Shu · Yao Dian* in Western Zhou Dynasty (1046 BC-771 BC). It said, "on Ri Zhong, you can find the constellation Niao in the center of the south, hence it is the middle of spring","on Ri Yong, you can find the constellation Huo in the center of the south, hence it is the middle of summer", "on Xiao Zhong, you can find the constellation Xu in the center of the south, hence it is the middle of autumn", "on Ri Duan, you can find the constellation of Mao in the center of the south, hence it is the middle of winter". "Ri" means day time, "Xiao" means night time, and "Zhong" means bisection. "Ri Zhong" and "Xiao Zhong" mean those two days when day time is equal to night time. That is the Spring Equinox and the Autumn Equinox. "Ri Yong" is the day when the day time is the longest, and that is the Summer Solstice. "Ri Duan" is the day when the day time is the shortest, and that is the Winter Solstice. 366 days is one cycle and form one year. That is to say, in Western Zhou Dynasty, our ancestors have formed the concept of the Spring Equinox, the Autumn Equinox, the Summer Solstice, the Winter Solstice by observing climate changes systematically, and made 366 days one year.

二、发展时期："分至启闭"与"八节"的出现

春秋战国时代，人们对天文及天气变化的认识进一步加强，开始从"两分两至"向"分至启闭"演变。《左传》中就多次谈到"分至启闭"，如《左传·僖公五年》载："凡分、至、启、闭，必书云物，为备故也。"其中，"分"指春分和秋分；"至"指夏至和冬至；"启"指立春和立夏，又指农作物发芽、抽叶、开花期；"闭"指立秋、立冬，也指农作物收获和保藏期。意思是说，但凡春分、秋分、夏至、冬至、立春、立夏、立秋、立冬八个节气，史官都要记录云气云色的气候变化，这是因为要为防止灾害做准备。这说明到春秋末期，二十四节气中重要的八节划分已经出现。但是，这八个节气的准确名称，则是在战国末期秦国丞相吕不韦主持编撰的《吕氏春秋·十二纪》中才出现的。至此，二十四节气的重要框架已经构成，而这"八节"的更迭，也基本反映了一年内自然界季节变化的过程。

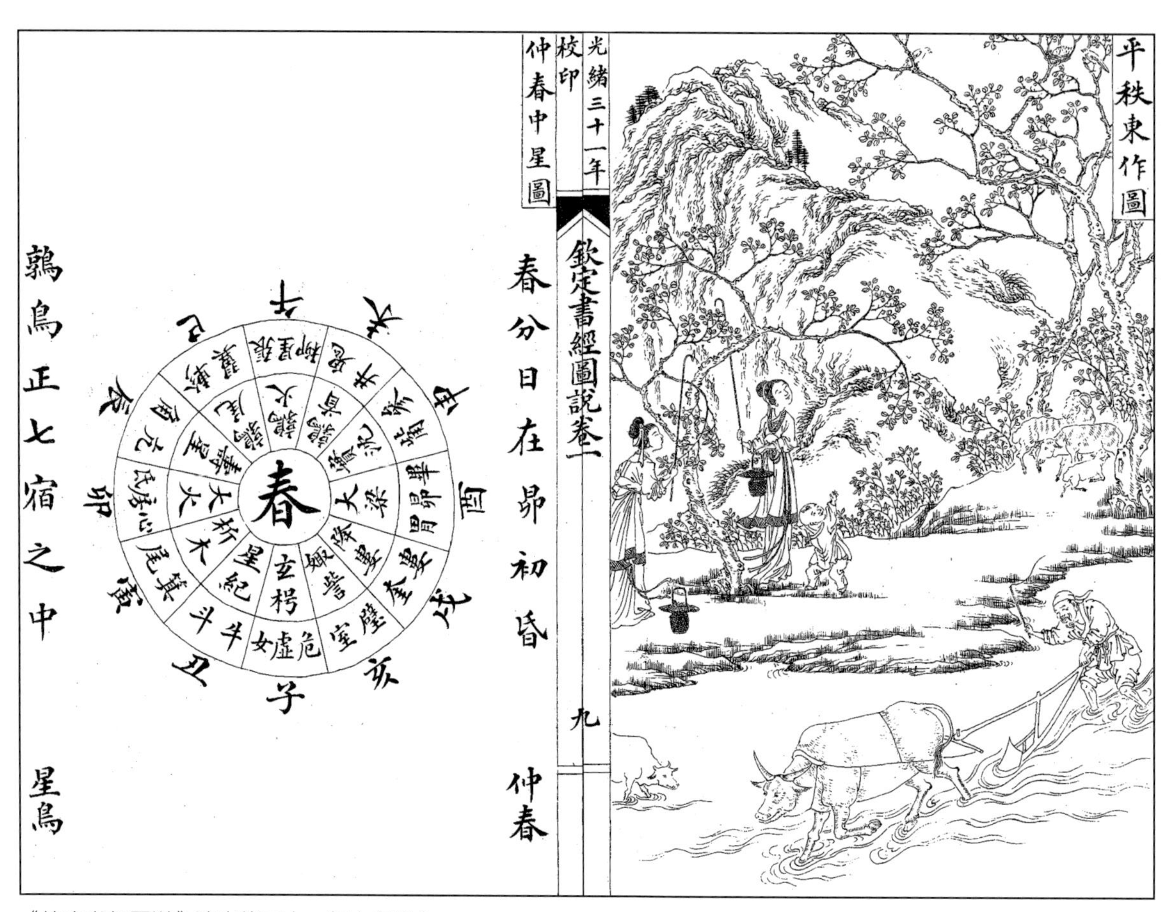

《钦定书经图说》清光绪三十一年内府刊本

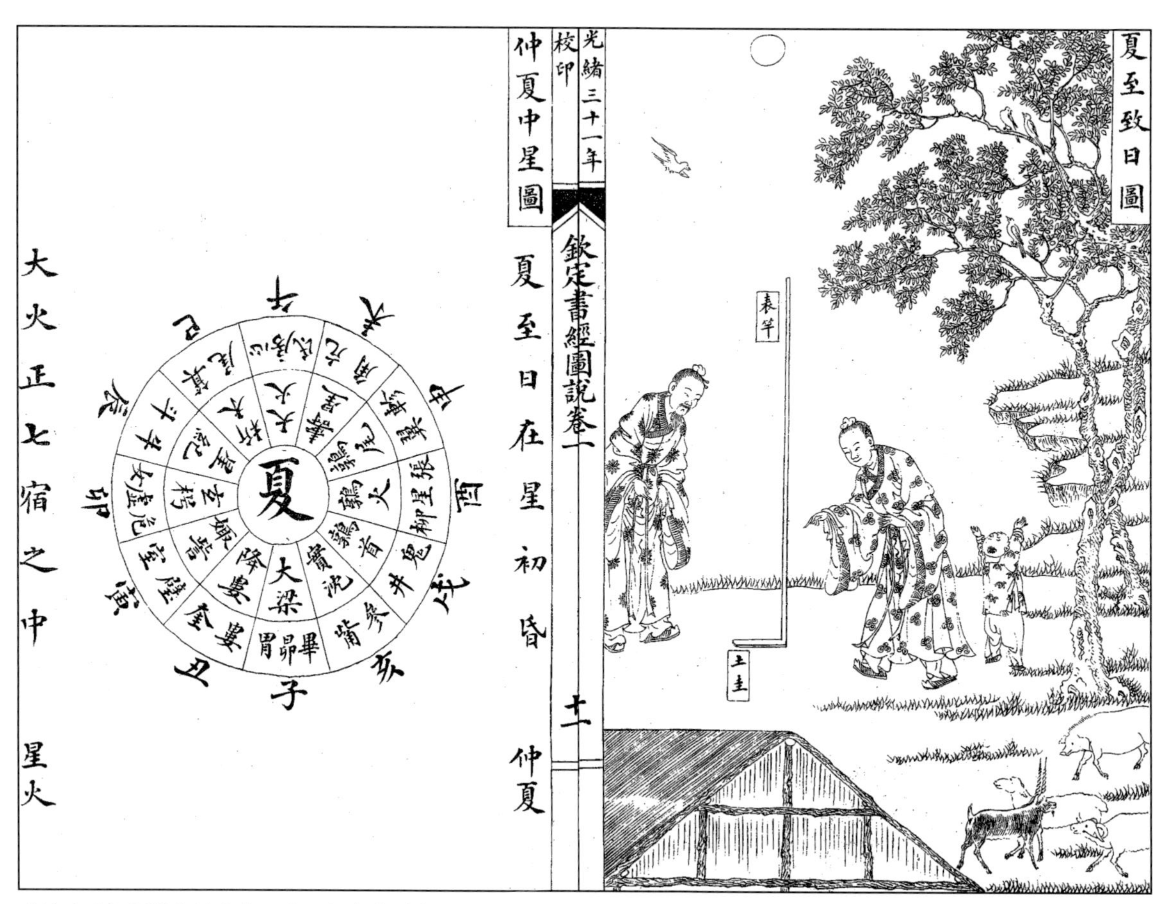

《钦定书经图说》清光绪三十一年内府刊本

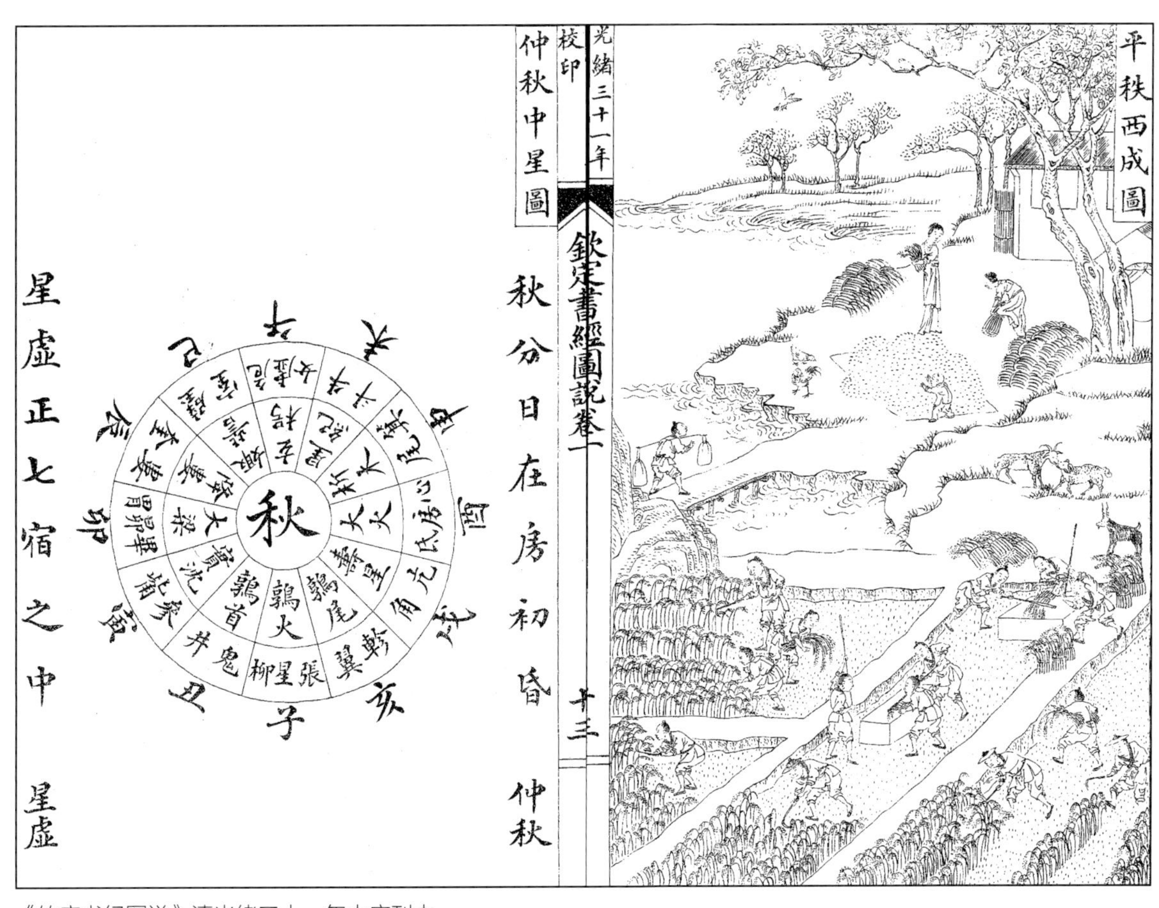

《钦定书经图说》清光绪三十一年内府刊本

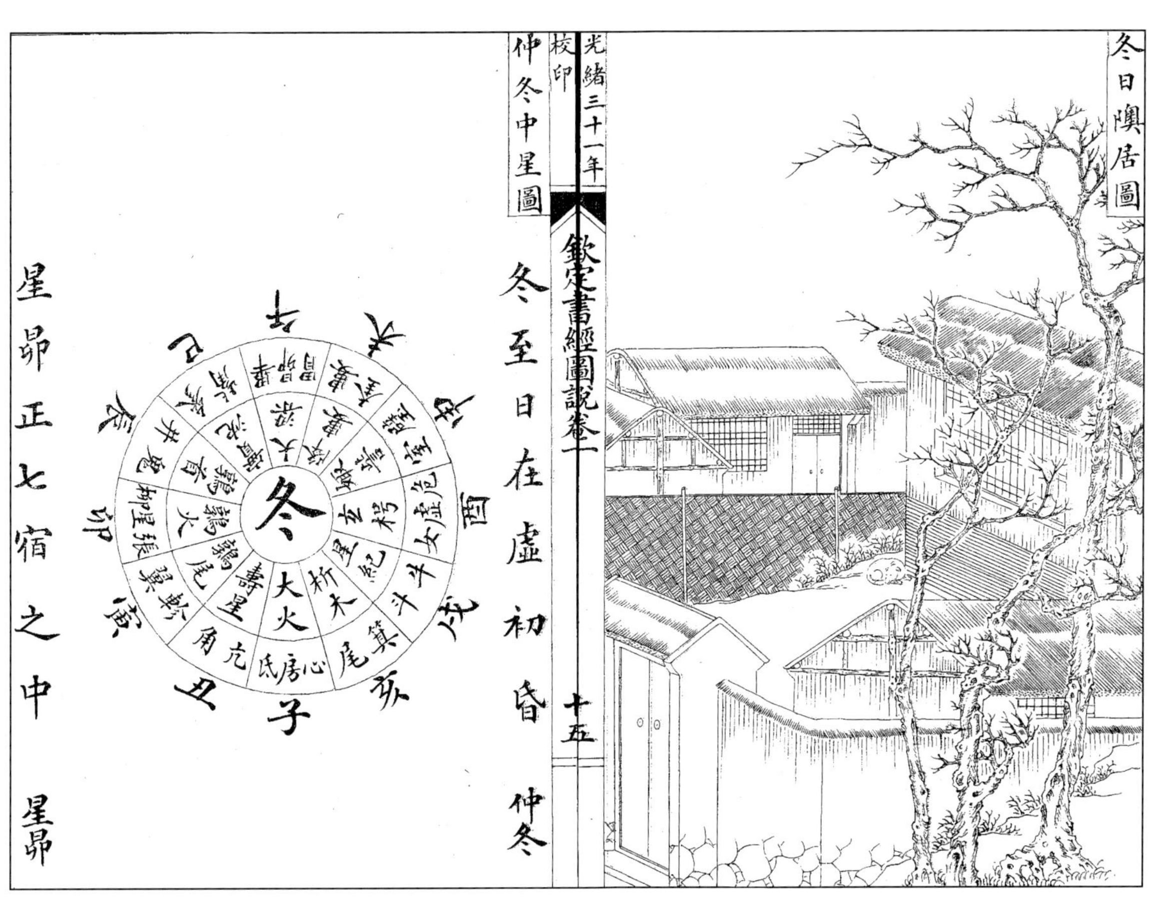

《钦定书经图说》清光绪三十一年内府刊本

Developing Stage: Emergence of "Fen Zhi Qi Bi" and 8 Solar Terms

During the Spring and Autumn Period and the Warring States Period, the knowledge of astronomy and climate change has been improved. "Liang Fen Liang Zhi" has evolved into "Fen Zhi Qi Bi". *The Chronicles of Zuo* has mentioned about "Fen Zhi Qi Bi" many times. *The Chronicles of Zuo•The Fifth Year of Duke Xi* has recorded that in the Spring Equinox, the Autumn Equinox, the Summer Solstice, the Winter Solstice, the Beginning of Spring, the Beginning of Summer, the Beginning of Autumn, the Beginning of Winter, official historian would record the change of cloud shape and color in order to make preparation for possible disasters. "Fen" means the Spring Equinox and the Autumn Equinox; "Zhi" means the Summer Solstice and the Winter Solstice; "Qi" means the Beginning of Spring and the Beginning of Summer, and also means the sprouting, leaf developing and blooming of crops; "Bi" means the Beginning of Autumn and the Beginning of Winter, and also means the harvest and store of crops. At the end of Spring and Autumn Period and the Warring States Period, the important 8 solar terms of 24 Solar Terms have appeared. The accurate name of the 8 solar terms have appeared in *Lv's Spring and Autumn Annals · Twelve Periods* compiled by Lv Buwei, prime minister of kingdom Qin at the end of the Warring States Period. Till then, the important framework of 24 Solar Terms has been formed. The changes of those 8 solar terms have basically reflected the seasonal change in a year.

三、定型完善时期："七十二候"的产生及"二十四节气"的正式颁布与实施

在节气发展的同时，"候应"也开始产生，并最早见于《吕氏春秋》。在该书《十二纪》中，除了八个节气名称的明确记载外，还分别在《孟春纪》提到"蛰虫始振"，《孟秋纪》提到"凉风至、白露降"，《季秋纪》提到"霜始降"等。这说明二十四节气在完成了对季节变化的划定后，又开始向着降水、降温和农作物生长等物候方面发展。完整的"七十二候"始见于《逸周书·时训解》。据专家研究，该书中物候的记述内容应晚于《吕氏春秋》，是在完整的二十四节气形成之后产生的。该书中把一年的时间大致按五日一候进行划分，并规定三候为一节，以与二十四节气相对应，每一候均以一种物候现象作为"候应"。与二十四节气多为天气变化不同，"七十二候"更多地利用了动植物以及大自然的多种变化作为时间标志，如"水始涸""地始冻""鸿雁来"等。"七十二候"以五日为一候，以"候应"为表征，划分更为细致，指导农时更为具体，形成了较为严格的全年物候历。

風暴雨數至藜莠蓬蒿竝興木仁金殺而行其令氣不和故民疫病也金生水與水相干故風雨數至荒穢滋生是以藜莠蓬蒿竝興○月令疾風作猋風數至作總至行冬令則水潦為敗霜雪大摯首種不入春陽冬陰也而行其令陰乘陽故水潦為敗雪霜大摯傷害五穀春為歲始稼穡應之不成熟也故曰首種不入○案月令注云舊說首種謂稷

孟春紀

一曰始生之者天也養成之者人也始初也能養天之所生而勿攖之謂之天子攖猶戾也○舊本作謂天子無之字孫據太平御覽七十七增天子之動也以全天為故者也全猶順也天性也故事也此官之所自立也官正也自從也立官者以全生也生性也今世之惑主主謂王也多官而反以害生則失所為立之矣多立官致任不肖人亂象干度故以害生也失其所為立官之法也譬之若修兵者以備寇也今修兵而反以自攻則亦

《吕氏春秋》（注）汉 · 高诱

Finalizing Stage: Emergence of 72 Phenological Terms and Fanalization of 24 Solar Terms.

With the development of the concept of solar terms, "phenological term" has appeared and was recorded in *Lv's Spring and Autumn Annals* for the first time. In the volume of *Twelve Periods*, "dormant insects begins to sing" was recorded in the *First Period of Spring*; "cool wind blows and white dew drops" was recorded in the *First Period of Autumn*; "frost starts to appear" was recorded in the *Last Period of Autumn*. After recognition of seasonal change by the 24 Solar Terms, the phenological terms which describe the temperature change and crops growth have started to form. The complete 72 phenological Terms have been found in the book *Yi Zhou Shu•Shi Xun Jie*. According to research, those records are later than *Lv's Spring and Autumn Annals*, and the emergence of phenological terms is after the finalization of the 24 Solar Terms. Every phenological term has one symbolic appearance. The 72 Phenological Terms has made time stamp with changes of living things and the nature, such as “rive starts to dry up”, “land starts to freeze”, and “wild goose has come back”. Five days are called one pentad. The whole year are strictly divided into 72 Phenological Terms. The method divides the time more in details and will guide the agricultural activities more specifically.

西汉初年，淮南王刘安的《淮南子》一书在《天文训》中对二十四节气进行了完整记载。在汉武帝太初元年（公元前104年），二十四节气在正式颁布的新历法——《太初历》中完整亮相。该历法由四川阆中人落下闳及邓平等人组织编制，除了正式把二十四节气定于历法之外，落下宏还使用他研究制作的“赤道式浑仪”测定二十八宿的赤道距度，并与二十四节气联系起来，明确了二十四节气的天文位置。东汉章帝建初七年（公元82年），在史学家班固写的《汉书·律历志》中，第一次将二十四节气作为历法载入史册。此后，隋朝天文学家刘焯又在汉代定型的二十四节气基础上，提出了“定气法”。在此之前，早期的二十四节气是将一周年平分为二十四等分来确定二十四节气的，从立春开始，每过15.22日就交替一个新的节气，这种做法被称为“平气法”。但是，太阳的周年视运动是不等速的，在各个平气之间，太阳在黄道上所走的度数并不相同。“定气法”则对此进行了完善，以太阳在黄道上的位置为标准，自春分点作为0° 起算，黄经每隔15° 为一个节气，二十四节气正好为360° ，由此使二十四节气的划分变得更为科学。不过，这一方法比较复杂，长久未能得到推广，直至清朝时期才在《时宪历》中正式采用。如今二十四节气的划分就是使用刘焯的“定气法”。

At the beginning of Western Han Dynasty, Liu An, king of Huai Nan has recorded 24 Solar Terms completely in Astronomy chapter of his book *Huai Nan Tzu*. In the first year of Taichu, Emperor Wu of Western Han Dynasty (104 BC), the 24 Solar Terms were fixed in new official calendar, *Taichu Calendar*. That calendar was complied by Luo Xiahong from Langzhong County, Sichuan Province. In addition to formally fixing the 24 Solar Terms in calendar, he has used his Equatorial Armillary Sphere to measure the equatorial distances among 28 mansions. He has connected the 24 Solar Terms with the distances and fixed the celestial location of solar terms. In the 7th year of Jianchu, Emperor Zhang of Eastern Han Dynasty (82 AD), historian Ban Gu has recorded the 24 Solar Terms in *History of Han Dynasty*, Calendar Chapter for the first time. Hereafter, Liu Zhuo, astronomer in Sui Dynasty (581-619), has pointed out “Ding Qi Fa”. In early stage, 24 Solar Terms have divided one year into 24 pieces equally. Starting from

《月令辑要》清康熙内府四色套印本

《农书》元·王祯著　明嘉靖山东布政司刊本

《月令广义》明万历秣陵陈邦泰刊本

《钦定书经图说》清光绪三十一年内府刊本

the Beginning of Spring, there will be one new solar term after 15.22 days. That way of fixing solar term is called "Ping Qi Fa". However, the yearly moving speed of the sun is not equal. The celestial longitude change between two solar terms is not the same. "Ding Qi Fa" has improved the measure. It make sun's ecliptic location the standard. The point of the Spring Equinox is 0° and there will be one solar term every other 15°. The whole 24 Solar Terms are just 360°. That measure is more scientific. However, that measure is too complicated. Therefore, it was officially used in *Shi Xian Calender* only in Qing Dynasty. "Ding Qi Fa" is the current way of dividing the 24 Solar Terms.

四、传承弘扬时期："二十四节气" 入选人类非物质文化遗产代表作名录

二十四节气体系成熟定型之后，中国古代先人长期按照节气的循环往复安排着自己的生产与生活，渐渐形成了与二十四节气相关的、丰富多彩的民俗文化。进入近现代以后，随着时代的发展和中外交流，中国在经济结构、科技水平以及生活方式等方面都有了许多变化，国际通用的公历已被现代中国人普遍使用，农业生产也更多地利用高科技手段来改变农作物的生长环境，以提高其产量和质量。可以说，二十四节气在现代中国社会生产和生活中的地位与作用正在发生变化。作为我国先民的文化创造，二十四节气已成为中华民族独特的文化时间与宝贵的文化遗产。为了更好地传承、弘扬这一文化遗产，2006年5月，二十四节气被批准列入第一批国家级非物质文化遗产名录；2016年11月，中国向联合国教科文组织提出申请并通过审议，"二十四节气——中国人通过观察太阳周年运动而形成的时间知识体系及其实践"，被列入联合国教科文组织人类非物质文化遗产代表作名录。从此，二十四节气的发展步入新的历史时期，包括节气菜的挖掘整理、传承创新及其他各种各样的活动不断涌现，进一步丰富着二十四节气的文化内涵，也推动了二十四节气的传承与弘扬。

Carrying Forward Stage: the 24 Solar Terms Has Been Inscribed on the Representative List of the Intangible Cultural Heritage of Humanity

Ancient Chinese have arranged life and production cyclically according to the 24 Solar Terms system, and developed colorful folk culture. After entering modern times, the economic structure, scientific and technological level, life style of China have changed significantly with the development of times and international exchange. Nowadays, Chinese people are using the Gregorian Calendar, and high tech is used to change the production environment of crops to increase quantity and quality. The position and function of the 24 Solar Terms in modern China are changing. However, it was created by our ancestors and has become a particular cultural time and precious cultural heritage of Chinese nation. In order to carry forward this heritage, the 24 Solar Terms have been inscribed on the Representative List of the National Intangible Cultural Heritage of Humanity in May 2006. In November 2016, the 24 Solar Terms have been inscribed on the Representative List of the Intangible Cultural Heritage of Humanity by UNESCO. Since then, the development of the 24 Solar Terms has stepped into a new era. In order to enrich the cultural connotations of the 24 Solar Terms and to carry it forward, there will be lots of activities, including sorting out the solar term dishes and carrying them forward.

二十四节气与川菜食材

The 24 Solar Terms and Food Ingredients of Sichuan Cuisine

自古以来，二十四节气就是华夏先人在农业生产、饮食起居乃至养身保健等方面基本遵循的时间法则。二十四节气客观反映了一年四季自然界不同节令的物候变化与动植物的生长周期，是中华民族农耕文明长期发展的结晶。二十四节气与食材生产的关系十分紧密，我国先民很早以前就总结出了“春生、夏长、秋收、冬藏”的食材生产规律。《夏小正》《礼记》《管子》《四民月令》《月令辑要》《淮南子》《齐民要术》《月令广义》等许多古代典籍，都强调了要遵循“四时”变化来从事农业生产和饮食起居活动。节气不同，其天气、物候、环境各异，食材的种植、养殖、成熟、收获及烹制加工等，也常常需要与不同的节气相适应，只有做到“天人相应”，才可能获得丰富多样的食材，从而满足社会发展和人体健康的饮食需求。

The 24 Solar Terms are the law of time which have been followed by Chinese people in agricultural production and dietary activities since the ancient times. It is the crystallization of Chinese farming civilization and reflect the change of phenology and growth circles of plants and animals in nature. The 24 Solar Terms are connected with food ingredients producing very closely. Our ancestors have found the rule of “getting born in spring, growing in summer, harvesting in autumn, storing up in winter” long time ago. Lots of famous ancient books, like *Xia Xiao Zheng*, *The Book of Rites*,

《月令广义》明万历秣陵陈邦泰刊本

《月令辑要》清康熙内府四色套印本

Guan Zi, *Si Yue Min Ling*, *Huai Nan Tzu*, *Qi Min Yao Shu*, have stressed the importance of carrying out agricultural production and dietary activities according to the change of seasons. The weather, phenology and environment are different in different solar terms. Food ingredients producing, harvesting and cooking should be compatible with different solar terms. Only when man corresponds to nature, people can get abundant and various food ingredients and meet the demand of health.

四川地区群山环抱、江河纵横、气候温和、雨量充沛、土地肥沃、四季常青。勤劳智慧的四川人民，自古以来就一直按照节气物候的发展变化来组织农业生产，民间也广泛流传着许多通俗易懂的节气农谚，如“立春雨水到，早起晚睡觉”，强调立春时节要及时农耕；“芒种不种，再种无用”，强调芒种时节要及时移栽水稻；“过了寒露，秋粮入库”，强调寒露之后要收获秋粮；“小雪铲白菜，大雪铲菠菜”，强调小雪、大雪时节要及时收获白菜、菠菜，等等。四川人按时令节气进行食材生产，不违农时、辛勤劳作，不仅形成了一些节气的食材生产习俗，也使得四川地区的食材品种丰富多样，禽畜河鲜、五谷杂粮、瓜果蔬菜应有尽有，为川菜烹饪奠定了坚实的物质基础。由于许多食材的生长周期较长，不是每一个节气都会产生至关重要的影响，因此，这里分别以春、夏、秋、冬“四时”为序，分别阐述各个季节一些重要节气中的川菜食材，尤其是季节性较强的蔬菜、水果的出产情况。

Sichuan region is surrounded by mountains. In this region, there are vertical and horizontal rivers, warm climate, abundant rainfall, fertile land and evergreen plants. Industrious and clever Sichuan people have organized agricultural production according to the changes of phenology since ancient times. There are lots of farmer’s proverbs in Sichuan. “Farmers should get up early and go to bed late in the Beginning of Spring and Rain Water” means people should start farming in time in the Beginning of Spring; “it will be useless if farmers do not sow seeds in Grain in Beard” means people should transplant rice in time in Grain in Beard; “people should store the grains harvested in autumn after Cold Dew” stresses the importance of harvest after Cold Dew; “harvest Chinese cabbage in Lesser Snow and harvest spinach in Greater Snow” stresses the importance of harvesting Chinese cabbage and spinach in time in Lesser Snow and Greater Snow. Sichuan people produce food ingredients according to solar terms. They follow farming seasons, work hard, invent customs of food ingredients producing, and make Sichuan food ingredients abundant and colorful. There are all kinds of poultry, river fish, cereals, vegetables and fruits, which have provided solid material foundation for the cooking of Sichuan cuisine. Not all the solar terms can play important role in producing food ingredients because the growth cycle is very long. Hence, we’d like to introduce the production of food ingredients for Sichuan cuisine, especially seasonal vegetables and fruits in some important solar terms, following the order of seasons.

一、春季节气与食材

春季包括立春、雨水、惊蛰、春分、清明、谷雨共六个节气。俗话说，“一年之计在于春”，一年中的主要农事活动大抵都是从此发轫。从立春开始，四川地区春回大地、万物复苏、草长莺飞，气温由10℃左右逐渐升高到22℃左右，降雨量逐渐增多，气候宜人，非常适合食材种植。“立春雨水到，早起晚睡觉”“过了惊蛰节，锄头不能歇”“清明下种，谷雨下秧”等民间农谚，都传递出同样一个信息——农耕的关键时节到了！与此相应，四川各地的人们都在抓紧时间进行食材的种植和养殖活动。此刻，资阳一些特色食材的种植和养殖也不失时机地应时而为，如小龙虾要及时撒下虾苗，安岳红薯要做好育秧工作，莲藕一般要在清明前后种植，紫竹姜要在谷雨时节播种，等等。

Spring Solar Terms and Food Ingredients

There are 6 solar terms in spring: the Beginning of Spring, Rain Water, the Waking of Insects, the Spring Equinox, Qingming, Grain Rain. As the proverb goes, “the whole year's work depends on a good start in spring”. The agricultural

activity will start in Spring. Since the Beginning of Spring, the spring has come back to Sichuan region and everything has come back to life. The temperature has risen from 10℃ to 22℃ gradually, and the rainfall is increasing. The climate is good for planting food ingredients. There are lots of farmer's proverbs there, like "farmers should get up early and go to bed late in the Beginning of Spring and Rain Water", "farmers can not rest after the Waking of Insects", "sow seed in Qingming, transplant rice seedlings in Grain Rain", etc. People all over Sichuan are seizing the time to plant and breed food ingredients. Some special food ingredients of Ziyang are planted and bred in this period. People will breed the baby crayfish, transplant the seedlings of Anyue sweet potato. Lotus root will be planted around Qingming, and Zizhu ginger will be sowed in Grain Rain.

在春季时节，除了食材的种植和养殖，四川地区还出产了许多季节性食材，可供川菜烹饪之用。从立春伊始，至清明、谷雨，四川许多越冬蔬菜开始源源不断地走上人们的餐桌，除了青菜头、棒菜、儿菜、瓢儿白、折耳根、韭黄等四川优质特产蔬菜外，还有莴笋、生菜、菠菜、芹菜、茼蒿、牛皮菜，以及春笋、三月瓜、嫩胡豆、新鲜豌豆等常见的应季蔬菜，品种繁多，琳琅满目。四川人将各种蔬菜以不同方式进行烹制加工，或凉拌、清炒，或制汤、做饼，虽然食法多样，但总体上都遵循着春季宜食"生发之物"的养生理念。同时，从立春到谷雨，柑橘类水果，如春见（粑粑柑）、沃柑、丑柑，以及草莓、樱桃等陆续成熟上市，人们在踏青之时，可采摘、品尝春天的新鲜水果，享受大自然馈赠的春季美味，其乐融融。

In spring, Sichuan people will harvest many seasonal food ingredients for cooking Sichuan dishes in addition to planting and breeding food ingredients. From the Beginning of Spring to Qingming, Grain Rain, many winter vegetables have been sent to the dinning table, like cabbage head, mustard, houttuynia cordata, hotbed chives, lettuce, asparagus lettuce, spinach, celery, chard, spring bamboo shoots, zucchini, lima bean, pea, etc. Sichuan people cook vegetables in many ways, like making cold dish in sauce, frying, making soup and making cake. Various ways of cooking follow the theory that people should eat developing foods in spring. Meanwhile, citrus like Chun Jian, Wo Gan, Chou Gan and strawberry, cherry are on the market from the Beginning of Spring to Grain Rain. People can pick and taste fresh fruits, and enjoy the delicacy of spring from nature when going for outing.

二、夏季节气与食材

夏季包括立夏、小满、芒种、夏至、小暑、大暑共六个节气。在四川地区，夏季气温一般在22℃～35℃，天气炎热、阳光炽烈、雷雨增多、湿气较重，万物郁郁葱葱、繁华茂盛，这是一个食材长势迅速的时节，农业生产进入了热火朝天的阶段。民间也流传着许多与之相应的农谚，如“多插立夏秧，谷子收满仓”“芒种不种，再种无用”“小暑连大暑，除草防涝莫踌躇”，等等。此时，人们一方面进行着秋熟食材的播种、移栽；另一方面则抓紧时间及时收获夏熟食材，同时还要加强田间管理，以确保各种食材健康成长。

Summer Solar Terms and Food Ingredients

There are 6 solar terms in summer: the Beginning of Summer, Lesser Fullness of Grain, Grain in Beard, the Summer Solstice, Lesser Heat, Greater Heat. The temperature in Summer is from 22℃ to 35℃ in Sichuan. It is hot, sunny, humid with lots of thunderstorm. The plants are green and arborous. It is a season for food ingredients to grow quickly. There are lots of farmer’s proverbs, like “transplant more rice seedlings in the Beginning of Summer, you can harvest abundant grains”, “it will be useless if farmers do not sow seeds in Grain in Beard”, “pay attention to weeding and preventing waterlogging in Lesser Heat and Greater Heat”, etc. On the one hand, people will sow and transplant autumn food ingredients; on the other hand, people will seize the time to harvest summer food ingredients and to strengthen fields managements for the healthy growth of food ingredients.

夏季时节，四川地区大量的春种蔬菜已成熟收获，许多水果已批量上市。从立夏、小满到小暑、大暑，四川各地逐渐成熟、出产的蔬菜层出不穷，既有辣椒、四季豆、豇豆、卷心菜、番茄、仔姜、黄花、空心菜、毛豆角，也有黄瓜、苦瓜、冬瓜、菱角、扁豆等，这些食材具有健脾开胃、清利湿热、祛暑益气的功效，非常适合夏季养生食用。同时，夏季是出产水果极多的季节，四川地区许多水果陆续挂满枝头，如桑葚、车厘子、枇杷、梅子、桃子、杏子、李子、西瓜、葡萄、荔枝，等等。这些水果既可生吃，又可制作菜肴，还可以做成饮品，在民间形成了芒种煮梅、大暑吃西瓜等饮食习俗。

In summer, many vegetables planted in spring can be harvested and many fruits are on the market. From the Beginning of Summer, Lesser Fullness of Grain to Lesser Heat, Greater Heat, there are too many vegetables good for harvest, like chilly, green bean, cowpea, cabbage, tomato, tender ginger, day lily, water spinach, edamame, cucumber, bitter gourd, winter melon, water caltrop, lentil, etc. Those food ingredients can strengthen spleen, promote appetite, dispel dampness and summer heat, tonify qi, and good to eat in summer. Meanwhile, summer is a fruit season. In Sichuan region, there are mulberry, cherry, loquat, plum, peach, apricot, watermelon, grape, lichi, etc. Those fruits can be eaten raw, made into dishes or drinks. There are eating customs of cooking plum in Grain in Beard and eating watermelon in Greater Heat.

这一时节，资阳地区也出产了许多特色食材。进入立夏、小满时节，果形硕大、色泽黄亮、甘甜多汁的雁江区伍隍枇杷已经成熟。作为四川地区枇杷中的上品，如今，伍隍枇杷已经产业化生产，规模大、品质优、产量高。而乐至县则进入到繁忙的采桑养蚕、喜摘桑葚的时节。民谚曰："勤喂猪，懒喂蚕，四十天，就见钱。"蚕蛹含有丰富的蛋白质和多种氨基酸，可用炸、炒、炖、卤、煮等烹饪方法制成蚕蛹系列菜肴；桑叶可凉拌、烹炒做成桑叶菜肴，又可制成桑叶面点，还可以制成桑叶茶；桑葚颜色紫亮，甘甜多汁，可以酿成桑葚酒。时至芒种、夏至之际，雁江区中和镇养殖规模超过万亩的小龙虾已逐渐长大，年产量更是以数千吨计。每逢此刻，当地举办的"小龙虾美食节"，都会吸引大量的食客蜂拥而至，而今已远近闻名。大暑时节，资阳荷花争相吐艳，荷叶接天连碧，在此期间隆重举办过多年的"丹山荷花节"，也会吸引大量的游客来此观赏荷花，品尝荷花、荷叶美食。

In this season, there are lots of special food ingredients in Ziyang reigon. In the Beginning of Summer and Lesser Fullness of Grain, Wuhuang loquat from Yanjiang District is on the market. The loquat is big, bright, sweet and juicy, which is the best of Sichuan loquat. Nowadays, Wuhuang loquat's production is industralized. In Lezhi County, people are busy with breeding silkworm and picking mulberry. As the proverb goes, "industrious people raise pigs, lazy people breed silkworm and can get profit in 40 days". There are abundant protein and amino acid in silkworm pupae. Local people will fry, stew, braise, boil silkworm pupae. Mulberry leaves can be made into cold dish or cooked. Local people will make mulberry leaves pastry, mulberry tea and mulberry wine. Crayfish in Zhonghe Town of Yanjing District has grown up in Grain in Beard and the Summer Solstice. There are more than ten thousand mu crayfish ponds in Zhonghe Town and the annual production is several thousand tons. Every year, there is a famous Crayfish Festival. In Greater Heat, lotus is blooming in Ziyang. There is a Danshan Lotus Festival. Lots of tourists will come to enjoy lotus blossom and taste fine foods made with lotus blossom and leaves.

三、秋季节气与食材

秋季包括立秋、处暑、白露、秋分、寒露、霜降共六个节气。在四川地区，秋季天气开始转凉，气温由30℃左右逐渐下降至16℃左右，日照时间逐渐减少，连绵阴雨天逐渐增多，湿气日重。正如农谚所说，“处暑满地黄，家家修廪仓”“白露白迷迷，秋分稻谷齐”“霜降一过百草枯，薯类收藏莫迟误”。此时，农家一方面要及时抢收秋熟作物；另一方面还要积极做好冬季食材的栽种工作，“晚种一天，少收一石”等民谚，就是对此时节非常形象的表述。

Autumn Solar Terms and Food Ingredients

There are 6 solar terms in autumn: the Beginning of Autumn, the End of Heat, White Dew, the Autumn Equinox, Cold Dew, Frost's Descent. It is getting cold in Sichuan in autumn. The temperature will go down from 30℃ to 16℃; the sunshine will be decreased gradually; it is rainy and humid. On the one hand, farmers will seize the time to harvest autumn crops, as the farmer's proverbs go, "the crops are ripe, and every family is building warehouse", "the dew makes grains

white in White Dew, and grains will be ripe in the Autumn Equinox", "all plants will wither after Frost's Descent, and yams must be harvested quickly". On the other hand, people should start planting the winter food ingredients, just as the farmer's proverb goes, "if you plant the corps one day late, your harvest will be one Dan less".

秋季时节，四川地区出产和收获的食材最为丰富，稻谷籽粒饱满，瓜果、蔬菜成熟飘香，许多禽畜、河鲜已秋膘贴身，肉质愈加鲜美。从立秋、处暑到寒露、霜降，四川的蔬菜品种依然十分丰富，除了持续上市的辣椒、茄子、四季豆、豇豆、扁豆、丝瓜、苦瓜、冬瓜、卷心菜、空心菜外，还有新鲜登场的花生、红薯、莲藕、板栗、银杏，等等。同时，果木枝头也是硕果累累，梨子、苹果、核桃、柑橘、柿子等众多果品已收获在即，尤其是经霜打后的水果，口味更鲜美、甘甜，汁水更丰富。此刻，伴随着菊花的绽放，螃蟹也进入到肥美体壮的最佳食用期。

In Autumn, harvest in Sichuan is the most abundant in a year. The rice is plump; the fruits and vegetables are fragrant; the poultry, livestock and river fish are fat and fresh. From the Beginning of Autumn, the End of Heat to White Dew, Frost's Descent, there are various vegetables on the market in Sichuan, like chilly, eggplant, green bean, cowpea, lentil, towel gourd, bitter gourd, cabbage, water spinach, peanut, sweet potato, lotus root, Chinese chestnut, ginkgo, etc. Meanwhile, fruits like pear, apple, walnut, citrus, persimmon are ripe. Fruits after frost are delicious and juicy. With the blossom of chrysanthemum, it is the best time to eat crab.

这一时节，也是资阳地区著名特色食材收获最多的时期之一。此时，拥有“国家地理标志保护产品”的安岳柠檬逐渐成熟上市。安岳柠檬含有丰富的柠檬酸和多种维生素、微量元素，对人体健康十分有益，是世界公认的美容保健水果。安岳柠檬仅在安岳县就种植了30余万亩、2 000多万株，产量约占全国总产量的80%。同时，资阳地区还用柠檬开发出了柠檬系列菜点及柠檬茶、柠檬汁等，对安岳获得“中国柠檬之都”、资阳获得“国际（柠檬）美食名城”起到了重要作用。此外，安岳红薯也是当地的著名特产，品种好、规模大、产值高，安岳

红薯也在此时进入收获季节。如今，安岳县的红薯加工产业逐渐壮大，已成为西南地区最大的红薯粉条生产加工基地。同样拥有“国家地理标志保护产品”的雁江蜜柑，是资阳市雁江区的著名特产。雁江蜜柑果实扁圆、光滑，富有弹性，具有成熟早、着色早、酸甜适度、果汁丰富、果肉化渣、风味浓、酸甜适度等特点，是柑橘中的上品。乐至莲藕作为当地的著名食材，也在此时进入收获季节。用乐至莲藕制成的“天池藕粉”闻名遐迩，既是“国家地理标志保护产品”和四川省名牌产品，也是藕粉中的佳品，最宜秋冬时节食用。

It is also the harvest time for famous food ingredients in Ziyang region in this season. Anyue lemon with the national geographical indication is on the market. There are abundant citric acid, vitamins and micro-element in Anyue lemon, which is good for health and recognized as beauty fruit. In Anyue County, lemon fields are about three hundred thousand Mu. There are twenty million lemon trees whose production can account for 80% of national output. Ziyang people make dishes, tea and juice with lemon. Anyue County has got the name of "Lemon Capital of China" and Ziyang City has got the name of "International Lemon Delicacy City". Anyue sweet potato is local speciality with high output. Anyue County is the biggest sweet potato noodle production base in southwest of China. Yanjiang Mi Gan (a kind of citrus) with the national geographical indication is good looking, early ripe, proper sweet and sour, juicy and delicious, which is among the best citrus in China. Lezhi lotus root is another local speciality. Tianchi lotus root flour made with Lezhi lotus root is Sichuan's famous-brand product with the national geographical indication. It is among the best and good to eat in autumn and winter.

四、冬季节气与食材

冬季包括立冬、小雪、大雪、冬至、小寒、大寒共六个节气。在四川地区，冬季气温由立冬时的16℃左右逐渐下降到0℃左右，虽然较为寒冷，但很难出现河水大量结冰的现象。随着日照减少，阴天增多，降雨量却相对较低。这一时节，四川与全国大部分地区相似，基本处于食材冬藏阶段，农事活动主要包括防寒收菜、追加冬肥、果树管理、兴修水利、疏松土壤等。

Winter Solar Terms and Food Ingredients

There are 6 solar terms in winter: the Beginning of Winter, Lesser Snow, Greater Snow, the Winter Solstice, Lesser Cold, Greater Cold. Although it is cold in Sichuan's winter, and the temperature will go down from about 16℃ in the Beginning of Winter to 0℃, the rivers are hardly frozen. Sunshine and rainfall will be decreased, and overcast days will be increased. Like other regions of China, Sichuan is in winter storing up stage. Farmers will mainly prevent cold, harvest vegetables, apply winter fertilizer, manage fruit trees, build water conservancy projects, loosen soil.

四川地区的冬季，相对于北方天寒地冻、万物肃杀而言，则显得温柔了许多，茫茫大地上仍然随处可见青翠的绿色与勃勃的生机。北宋著名文学家苏轼曾有诗形象地描述了他故乡的场景："北方苦寒今未已，雪底波棱如铁甲。岂如吾蜀富冬蔬，霜叶露牙寒更茁。"即使是在寒气逼人的冬季，四川地区仍然出产了许多新鲜蔬菜，除了生菜、荸荠、蒜苗、莴笋、白菜、卷心菜、菠菜、芹菜等常见蔬菜外，更有色碧质嫩的豌豆尖、冬寒菜等四川特色优质蔬菜，点缀着四川人的餐桌。此外，圆润硕大的柚子以及冬草莓等水果也开始陆续上市。进入冬季，猪、牛、羊、鸡、鸭、鹅等禽畜体内的能量存储已达到顶峰，肉质十分肥美，为人们抵御寒冷，过一个祥和的春节增添了应有的物质基础。

Sichuan's winter is softer than the winter in the northern area. You can still find the green and vitality in this region. In winter, there are still lots of fresh vegetables on the market in Sichuan, like romaine lettuce, chufa, garlic bolt, lettuce, Chinese cabbage, cabbage, spinach, celery, tine pea, Chinese mallow. Moreover, fruits like big pomelo and strawberry are on the market too. In this season, pig, ox, goat, chicken and duck are delicious, which have provided material foundation to prevent coldness and to celebrate Spring Festival.

这一时节，资阳地区继续出产了许多著名的特色食材，拥有"国家地理标志保护产品"的安岳"通贤柚"已成熟收获。"通贤柚"果肉晶莹剔透、无核，果实香味浓郁、甜酸适度、脆嫩化渣、汁多爽口，深受人们喜爱。如今，通贤柚已成为当地民众的致富果。冬至、小寒时节，在四川许多地方的大街小巷，都可见烹煮羊肉的袅袅炊烟，乐至黑山羊肉就是其中的重要角色。乐至黑山羊通体黑色、被毛油亮、体型健壮、肉色红润、质地细嫩、香而不腻，极具山羊肉特有的香味。乐至县被誉为"中国黑山羊之乡"，产值高达11.53亿元，约占全县畜牧业总产值的30%以上。

There are also lots of special food ingredients in Ziyan in this season. Tongxian pomelo in Anyue County with the national geographical indication is ripe and good for harvest. That pomelo is good looking, seedless, fragrant, juicy, proper sweet and sour. It is very popular among people. Nowadays, planting Tongxian pomelo is a way of making money for local people. In the Winter Solstice and Lesser Cold, every family in Sichuan will cook mutton. Lezhi black goat meat is a very important food ingredients. Lezhi black goat is big shaped and covered with bright black wool. Its meat looks red and tastes delicate with a special fragrance. Lezhi County is called "Hometown of Chinese Black Goat". The production value of goat is 1.153 billion RMB, which accounts for 30% of the livestock industry output of the whole county.

二十四节气与食材生产密切相关，四川人始终遵循一年四季节令物候的变化与动植物的生长规律进行种植、养殖与收获，辛勤耕耘，劳作不辍，由此使得川菜拥有了类别丰富、品种繁多的食材基础。

In brief, the 24 Solar Terms are connected with food ingredients producing closely. Sichuan people plant, raise and harvest food ingredients according to the phenology changes and growth law of plants and animals. It has provided various and abundant material foundation to Sichuan cuisine.

二十四节气与川菜养生

The 24 Solar Terms and Regimen of Sichuan Cuisine

所谓“养生”，是指通过各种方法增强自身体质、预防疾病，从而达到以保养身体、健康长寿为目的的一种活动。“养生”一词始见于《庄子・养生主》：文惠君说“吾闻庖丁之言，得养生焉”。文惠君在听了庖丁叙述解牛的技巧后悟出只有遵循自然之道才能养生。现存最早的中医学理论著作《黄帝内经》指出，适应季节寒温的变化、顺应自然阴阳的变化和调节情绪是养生的关键。

Regimen is called “Yang Sheng” in Mandarin which means that people strengthen physique and prevent disease through all kinds of methods, and the purpose is to nourish body and prolong life. The word “Yang Sheng” has been found in *Zhuangzi · Essentials of Nourishing Life*. Lord Wenhui said, “I have heard Pao Ding’s words and got the essentials of Yang Sheng.” After hearing the introduction on dismembering ox by Pao Ding, Lord Wenhui has known that the essential of Yang Sheng is following the natural law. *The Inner Canon of Huangdi*, the earliest existing book on TCM theory, has pointed out that the key of Yang Sheng is following the change of nature and yin-yang, and regulating emotion.

人类生活在自然界之中，是自然界的一部分，与自然有相通、相应的关系，因而必须遵循同样的运动变化规律。自然环境中存在着人类赖以生存的必要条件，人类需要摄取饮食，呼吸空气与大自然进行物质交换，从而维持正常的新陈代谢活动。同时，自然界的变化又直接或间接地影响着人体，促使人的机体相应地发生生理或病理反应，故《黄帝内经・灵枢》说“人与天地相应”，意谓人与天地是互相参照的，人体的生命活动与自然界息息相关。具体到饮食上，“天人相应”的直接表现是因地、因时、因人施膳。《黄帝内经・素问》对不同地区人的外貌、体质、口味等特点已有论述，举例说：南方者，“其民皆致理而赤色，其病挛痹”；中央者，“其民食杂而不劳，故其病多痿厥寒热”。因南方雾露缭绕，当地人喜吃酸腐食物；中央地形平坦、多潮湿，当地食物种类较多，生活舒适，容易导致痿软等疾病。由此说明地理环境及气候等差异，对人体生理乃至疾病的发生都会有影响，正如俗语所言，“一方水土养一方人”，因此，养生或健身必须因地、因时制宜。就四川地区而言，其独特的地理、气候等条件，也促使川菜形成了自己的饮食养生特点。结合四川地区的气候特征、地理环境及物产资源等因素，可大致梳理出不同季节及节气与川菜养生之间的关系。

類經一卷 攝生類 十三
者沉於下腎氣不蓄藏
則注泄沉寒等病生矣夫四時陰陽者萬物之
根本也生成之所由也所以聖人春夏養陽秋冬養陰
以從其根夫陰根於陽陽根於陰陰以陽生陽以陰長所以聖人春夏則養陽以爲
秋冬之地秋冬則養陰以爲春夏之地皆所以從其根也今人有春夏不能養陽者每因風凉
生冷傷此陽氣以致秋冬多患瘧瀉此陰勝之爲病也有秋冬不能養陰者每因縱慾過熱傷
此陰氣以致春夏多患火證此陽勝之爲病也善養生者宜切佩之故與萬物沉
浮於生長之門逆其根則伐其本壞其眞矣能順
陰陽之性則能沉浮於生長之門矣萬物有所生而獨知守其根百事有所由而獨知守其門

則聖人之能事也故陰陽四時者萬物之終始也死生
之本也陰陽之理陽爲始陰爲終四時之序春爲始冬爲終死生之道分言之則得其
陽者生得其陰者死合言之則陰陽和者生陰陽離者死故爲萬物之始終死生之本也逆
之則災害生從之則苛疾不起是謂得道苛音呵殘
虐也道者聖人行之愚者佩之聖人與道無違故能行之愚者信道
不篤故但佩服而已夫既佩之已匪無悟而尚稱爲愚令有并陰陽不知而曰醫者又何如其
人哉老子曰上士聞道勤而行之中士聞道若存若亡下士聞道大笑之不笑不足以爲道正
此謂也從陰陽則生逆之則死從之則治逆之則
類經一卷 攝生類 十四

《类经》明 · 张景岳著

Human beings live in nature and are a part of nature. We have been connected with nature and correspond to nature, and must follow the same law of change. Nature has provided necessities for human survival. People need take foods and drinks, and breathe air to maintain normal metabolism. Meanwhile, the changes of nature will affect human body directly or indirectly, and cause relevant physiologic and pathological reflections. *Inner Canon of Huangdi · Ling Shu* said "correspondence between man and universe", Which means taking foods according to place, time and individual. *Inner Canon of Huangdi · Su Wen* has different description on people's appearance, physique and taste in different regions. For example, in the southern region, people prefer to eat sour food. The central region is plain and humid; people there have many options on food and comfortable life, and easily suffer from body atrophies. It means that the difference of geography, environment and climate will affect physiology and the occurrence of diseases. Hence, regimen must depend on location and time. "Soil and water to support one party people". In Sichuan region, special geography and climate create the characteristics of Sichuan cuisine on regimen. Here, we will introduce regimen of Sichuan cuisine in different seasons and solar terms, on the base of Sichuan's climate and geographical environment.

一、节气与饮食养生的关系

《黄帝内经 · 素问》指出："人以天地之气生，四时之法成。"人不仅要依赖天地之气所提供的物质条件而生存，更要适应四季阴阳变化的规律才能发育成长。自然界的万物随着四时的气候变化，有春生、夏长、秋收、冬藏等生长变化过程与规律。人类也不例外，随着四时气候变化，人体在生理上也会有着相应的反应。比如，夏季时暑热盛行，阳气趋于体表，腠理（毛孔）开泄，肌体多出汗水以泄体热；而冬季寒冷当道，阳气趋于里，腠理闭以保温，多余水分变成尿液排出，这些表现，都是人体在不同气候下为适应自然而进行自我调节的结果。但夏季时人体大量出汗、水分流失太多，如不及时补充，则人体机能必将受到损伤，因此必须多饮水，并且进食相对清淡的饮食，方能达到养生的目的；而在冬季时，肌体为抵御寒冷，则应进食醇浓滋补之物，方为养生正途。可以说，四时节气与人体的饮食养生有着十分紧密的关系，在不同季节、不同节气中，人们的日常饮食应有所侧重，尽量与自然节律相吻合，才有可能满足人体正常的生理需求，从而实现养生强体的愿望。

Relations Between Solar Terms and Dietary Regimen

Inner Canon of Huangdi · Su Wen has pointed out that people's survival depends on materials provided by qi from the universe, and people's growth needs to follow the law of seasonal changes and yin-yang changes. All things on earth change according to the seasonal climate changes. There is the rule of "getting born in spring, growing in summer, harvesting in autumn, storing up in winter". Seasonal climate changes have reflections on human body. In hot summer, yang qi has gathered on surface, and pores puff more to sweat more. In cold winter, yang qi has hidden inside, and muscular interstices get to tighten to keep warm. Extra liquid has become urine and been excreted. That is the result of human body's self regulating to adapt different climates. However, in summer people should drink more water and eat light food to supplement the lost water because of too much sweating. In winter people should take more tonic foods to prevent coldness. So to speak, seasons and climates have close relations with dietary regimen. In order to meet the physiological need and nourish the body, there should be different focuses on diet in different seasons and solar terms.

二、节气与饮食养生的原则

（一）注重时令

吃时令菜是中华民族饮食养生的经验总结。孔子《论语》言："不时，不食。"饮食要应时令、按季节，不是当季节的食材则不吃。《黄帝内经》也提出"食岁谷"，即多吃时令谷物、果菜。清代袁枚的《随园食单》专门列有"时节须知"，指出"冬宜食牛羊，移之于夏，非其时也"，"辅佐之物，夏宜用芥末，冬宜用胡椒"。中国传统医学和养生学均认为"医食相通"，食物与药物一样具有"气""味"（即食性、药性），只有在当季节、合时令的情况下，即生长、成熟符合四时节气时，才能获得天地之精气，使其气、其味处于最佳状态。传统中医的阴阳气化理论认为，动植物都有一定的生长节律，必须经过一定的生长周期后才能成熟，气与味才能充足。如果违背自然生长周期，打破"春生、夏长、秋收、冬藏"的寒热消长规律，就会使食物寒热不调、气味混乱、品质下降，对人体的养生作用也会大大降低，正如袁枚在《随园食单·须知单》中所言："有过时而不可吃者，萝卜过时则心空，山笋过时则味苦，刀鲚过时则骨硬。所谓四时之序，成功者退，精华已竭，褰裳去之也。"

Principles of Solar Terms and Dietary Regimen

1. Paying Attention to Season

Eating seasonal dishes is a summarized experience in Chinese dietary regimen. *The Analects of Confucius* said, "do not eat in wrong time". Diet should be arranged according to seasons. People should not eat non-seasonal food ingredients. *Inner Canon of Huangdi* has pointed out "Shi Sui Gu", which means eating more seasonal grain, fruits and vegetables. Yuan Mei in Qing Dynasty has added one chapter of "Notice of Seasons" in his *Menu of Sui Yuan*. He pointed out that, "eating mutton and beef is good for winter but not good for summer", "it is good to take mustard in summer and pepper in winter". TCM and regimen believe that "medicine and food are interlinked". Both food and medicine have "qi" and "taste" (property of a medicine and food). Only in a proper season and time (the growth has followed the rule of season change), those medicine and food can get the essence of nature, and in perfect founction. According to the theory of yin-yang and qi, animals and plants have their own growth circle. They can get enough "qi" and "taste" after certain growth circle. If we disobey the circle and the rule of "getting born in spring, growing in summer, harvesting in autumn, storing up in winter", the quality, taste, inner balance between cold and hot of the food will be damaged, its function for regimen will be reduced too. Yuan Mei has given the examples. In his *Menu of Sui Yuan · Notice*, he said, "non-seasonal radish is hollow; non-seasonal bamboo shoot is bitter; non-seasonal coilia ectenes is hard. People should not eat non-seasonal foods because their essences have been depleted. "

在四川地区，一年四季瓜果更替井然有序，早在东晋时的《华阳国志·蜀志》就有"其山林泽渔，园囿瓜果，四节代熟，靡不有焉"之语，由此可见，蜀地丰富的食材来源，为制作时令菜提供了可靠的物质基础。因此，川菜烹饪十分注重四时节气，菜点的季节特色尤为突出，以此获得最佳的口感与养生效果。如回锅肉尤以冬春之季鲜嫩、碧绿的蒜苗炒制为佳；水煮肉片中的豆芽、莴笋追求新嫩爽脆，最佳时节是早春之际；秋冬之时，红薯味道香甜，用之垫底，制作粉蒸肉则吸足汁水、抢去主角猪肉的光环。

In Sichuan, there are various and abundant fruits in four seasons. *Chronicles of Huayang Kingdom · Shu* in Eastern Jin Dynasty has recorded, "in forest and rivers, in fields and orchards, there are abundant products in every season". Abundant resources have provided material foundation for making seasonal dishes. Hence, Sichuan cuisine pays attention to seasons, and Sichuan dishes which pursuit the best taste and function of regimen have distinct seasonal characteristics. For example, Double-Cooked Pork Slices needs tender green garlic bolt picked in winter and spring; Poached Spicy Pork Slices needs fresh tasty bean sprout and lettuce picked in early spring; Steamed Pork with Rice Flour needs sweet yam in autumn and winter, which will absorb the juice and get the flavor of meat.

中而見其真矣。以天地為之陰陽。此重申上文，言賢人之養身，皆法乎天地之陰陽。如天氣地氣風雷谷雨川海九竅之類皆是也。陽之汗以天地之雨名之。汗出於陽而本於陰，故以天地之雨名之。雨即人之汗，汗即天之雨，皆陰精之所化。知雨之為義，則可與言汗矣。陽之氣以天地之疾風名之。氣本屬陽，陽勝則氣急，故以天地之疾風名之。知陰陽之權衡，動靜之承制，則可與言氣矣。暴氣象雷。天有雷霆，火鬱之發也。人有剛暴，怒氣之逆也。故語曰雷霆之怒。逆氣象陽。天地之氣，升降和則不逆矣。天不降，地不升，則陽亢於上。人之氣逆，亦猶此也。故治不法天之紀，不用地之理，則災害至矣。上文

言人之陰陽，無不合乎天地，故賢人者必法天以治身。設不知此，而反天之紀，逆地之理，則災害至矣。此理字與前五里之里不同，蓋彼言廣輿之里，此言理氣之理。

陰陽之中復有陰陽 素問金匱真言論〇五

故曰：陰中有陰，陽中有陽。故曰，引辭也。既言陰矣，而陰中又有陰；既言陽矣，而陽中又有陽。此陰陽之道所以無窮，有如下文云者。平旦至日中，天之陽，陽中之陽也；日中至黃昏，天之陽，陽中之陰也；合夜至鷄鳴，天之陰，陰中之陰也；鷄鳴至平旦，天之陰，陰中之陽也。一日之氣，自卯時日出地上為晝，天之陽

《类经》明 · 张景岳著

（二）阴阳协调

《黄帝内经·素问》指出，阴阳四季的变化是万物盛衰存亡的关键和万物生命的根本，必须顺从这个规律，否则将会对生命造成伤害，因此，调整阴阳以适应自然是养生之根本，并且指出了四时养生的基本原则是“春夏养阳，秋冬养阴”。春季需要适应自然阳气的生发而养阳，饮食宜辛温发散；夏季在清解暑热的同时，尤应注意不可在饮食上太过寒凉，以免损伤阳气；而秋冬则需顺应自然界阳气的收敛、潜藏并顾护阴液，可多食具有滋阴、养肾作用的食物。

2. Coordination of Yin and Yang

Inner Canon of Huangdi · Su Wen has pointed out that the changes of yin-yang and four seasons are the key for all things to rise and fall, and the foundation of all things. We must follow this rule, otherwise our lives will be harmed. Therefore, adjusting yin-yang to adapt the nature is the basic of regimen. The basic principle of regimen in four seasons is “nourish yang in spring and summer, and nourish yin in autumn and winter”. In spring, we should nourish yang to adapt the natural growth of yang qi; in summer, we should avoid too much coldness which will damage yang qi when dispelling summer heat; in autumn and winter, we should follow the constriction of yang qi and protect the yin. We should eat more food which can tonify yin and nourish kidney.

中国传统医学和养生学认为，人体的阳气就像天空与太阳，对人体非常重要，正如《黄帝内经·素问》所言“阳气者，若天与日”。晒太阳是增补阳气的重要途径之一。但四川盆地常年多云雾，与国内其他地区相比，日照时数总体偏少，尤其秋冬季的日照更少，容易导致阳气不足而怕冷，人们便常常通过饮食来补充阳气。乾隆年间《夔州府志·风俗》提到川东地区“春夏多煖（即‘暖’），秋冬少寒，有山风气，辣、茶以避之”，民众在日常饮食中多吃辣以驱寒。四川人自古以来形成“好辛香”的饮食传统，从养生角度上讲，恰巧能养阳祛寒，无论是春季“五辛盘”与春饼中的韭菜、葱，还是夏季火锅中的辣椒、花椒，均为辛温之品，辛香可发散生阳，温能顾护阳气；冬至吃羊肉也是蜀人通过饮食补阳祛寒的养生法宝。此外，阴在人体，即为阴液，由于四川地区秋冬阴雨较多，相对不甚干燥，饮食只要不过于燥热，适量食用银耳、梨等食材即可达到养阴的目的。

TCM and regimen believe that the yang qi in human body is as import as sky and sun. Just as *Inner Canon of Huangdi · Su Wen* said, “yang qi is like sky and sun.” Sunbathe is an important way to supplement yang qi. However, Sichuan Baisn is cloudy and foggy. Its sunshine duration is among the least in China, espcially in autumn and winter, which will cause the deficiency of yang qi. Therefore, people need supplement yang qi through diet. *Kuizhou Local Chronicles · Custom* in Qianlong Period, Qing Dynasty mentioned that “the eastern Sichuan reigon is warm in spring and summer and cold with wind in autumn and winter, and people eat spicy food and drinking tea to protect themselves.” People normally eat spicy food to dispel coldness. Sichuan people’s dietary tradition of “preferring to pungent food can tonify yang and dispel coldness. Wu Xin Pan in spring, chives and green onion in spring cake, chilly and Sichuan pepper in hot pot are pungent food ingredients. They can help the growth of yang, keep warm and protect yang qi. Eating mutton in the Winter Solstice is a traditional way for Sichuan people to nourish yang and dispel coldness through diet. Moreover, yin in human body is yin liquid. It is rainy in autumn and winter, hence foods in those seasons can not be too dry and heat. People can eat proper tremella and pear to tonify yin.

（三）五味调和

1. 五味与四季的配合

五味最初是指“辛、甘、酸、苦、咸”五种滋味，后来代指具有相应作用的食物。《黄帝内经·素问》认为，五味调和得宜，会使骨骼端正、筋脉柔软、气血流畅、腠理致密，只要严格依照养生的方法就能长寿。《黄帝内经·灵枢》的“五味篇”，则指出了五味与人体之间的关系：“五味入于口也，各有所走，各有所病。”意即饮食进入人体之后，五味会分别进入各自对应的脏腑经络，而脏腑经络在五味的影响下也会产生各自的疾病。

3. Reconciling Five Flavors

Coordination Between Five Flavors and Four Seasons

At the beginning, five flavors mean pungent, sweet, sour, bitter, and salty. Later, they mean foods with those flavors. *Inner Canon of Huangdi · Su Wen* believes that if five flavors have been reconciled properly, human bones will be in proper way, tendon will be soft, qi and blood will be clear and skin will be delicate. People can prolong life by following the regimen strictly. The chapter of “Five Flavors” in *Inner Canon of Huangdi · Ling Shu* pointed out the relations between Five flavors and human body. After eating, five flavors will go into different organs and meridians, and organs and meridians will suffer from different diseases under the influence of five flavors.

不同的季节有不同的食物，也有应当侧重的味道及作用。《礼记·内则》言：“凡和，春多酸，夏多苦，秋多辛，冬多咸，调以滑甘。”即从五行的角度将季节与五味进行联系后指出：春宜养肝，酸入肝，酸味可养肝；夏宜护心，苦入心，苦味可护心；秋宜滋肺，辛入肺，辛味滋肺；冬宜益肾，咸入肾，咸味益肾。但是各味食用太过，则不仅容易伤及本脏，还可影响其他脏腑。因此，《千金方》为此提出了相应的食治原则，以避免出现这样的问题：“春省酸增甘以养脾气，夏省苦增辛以养肺气，长夏（相当于夏历六月）省甘增咸以养肾气，秋省辛增酸以养肝气，冬省咸增苦以养心气。”即春季宜减少酸味，增加甘味以养脾；夏季应减少苦味，增加辛味以养肺；长夏应减少甘味，增加咸味以养肾；秋季应减少辛味，增加酸味以养肝；冬季应减少咸味，增加苦味以养心。

黄帝曰願聞穀氣有五味其入五藏分別柰何伯高曰胃者五藏六府之海也水穀皆入于胃五藏六府皆禀氣于胃五味各走其所喜穀味酸先走肝穀味苦先走心穀味甘先走脾穀味辛先走肺穀味鹹先走腎穀氣津液已行營衛大通乃化糟粕以次傳下黄帝曰營衛之行柰何伯高曰穀始入于胃其精微者先出于胃之兩焦以溉五藏別出兩行營衛之道其大氣之摶而不行者積于胸中命曰氣海出於肺循喉咽故呼則出吸則入天地之精氣其大數常出三入一故穀不入半日則氣衰一日則氣少矣黄帝曰穀之五味可得聞乎伯高曰請盡言之五穀秔音庚米甘麻酸大豆鹹麥苦黄黍

靈樞卷八

辛五果棗甘李酸栗鹹杏苦桃辛五畜牛甘犬酸猪鹹羊苦雞辛五菜葵甘韭酸藿鹹薤苦葱辛五色黄色宜甘青色宜酸黑色宜鹹赤色宜苦白色宜辛凡此五者各有所宜五宜所言五色者脾病者宜食秔米飯牛肉棗葵心病者宜食麥羊肉杏薤腎病者宜食大豆黄卷猪肉栗藿肝病者宜食麻犬肉李韭肺病者宜食黄黍雞肉桃葱五禁肝病禁辛心病禁鹹脾病禁酸腎病禁甘肺病禁苦肝色青宜食甘秔米飯牛肉棗葵皆甘心色赤宜食酸犬肉麻李韭皆酸脾色黄宜食鹹大豆豕肉栗藿皆鹹肺色白宜食苦麥羊肉杏薤皆苦腎色黑宜食辛黄黍雞肉桃葱皆辛

靈樞卷八終

《黄帝内经·灵枢》明嘉靖时期刊本

There are different foods in different seasons. The tastes and functions are different too. *The Book of Rites · Neize* said, "It's harmonious to have more sour in spring, more bitter in summer, more pungent in autumn and more salty in winter". It connected seasons with five flavors from the perspective of five elements. It is good to tonify liver in spring because sour flavor can enter the liver and tonify it. It is good to protect heart in summer because bitter flavor can enter heart and protect it. It is good to moisten lung because pungent flavor can enter lung and moisten it. It is good to benefit kidney in winter because salty flavor can enter kidney and benefit it. However, take too much five flavors will not only hurt its own organs but also others. Therefore, *Thousand Golden Prescriptions* has mentioned that people can avoid those problems by food dietary therapy. In spring, people should reduce sour flavor and increase sweet flavor to nourish spleen. In summer, people should reduce bitter flavor and increase pungent flavor to nourish lung. In long summer, people should reduce sweet flavor and increase salty flavor to nourish kidney. In autumn, people should reduce pungent flavor and increase sour flavor to nourish liver. In winter, people should reduce salty flavor and increase bitter flavor to nourish heart.

2.五味与地区气候的适应

四川盆地的地势西北高，东南低，气候温暖湿润，冬无严寒，无霜期长。其中的川西平原植被丰富，气候温暖湿润，生态环境良好；川东地区多为丘陵山区，气候常阴冷潮湿，山林之间多瘴气，人易患病，从而促使四川人比较注重养生、防病。与其他地区相比，四川盆地每月的平均相对湿度均保持在70%～80%，四季湿度较大。四川部分市、州的降雨量如下表所示。

Adaptation of Five Flavors and Climate

The terrain of Sichuan Basin is high in the northwest and low in the southeast. It is warm and humid. The winter is not severely cold and the frost-free season is long. The Western Sichuan Plain has rich plants, warm and humid climate, good ecological environment. The eastern Sichuan region is hilly, cold and humid, full of miasma which will cause disease easily. Hence, Sichuan people pay attention to regimen and preventing diseases. Monthly average relative humidity in Sichuan Basin is from 70% to 80%, and the humidity is high comparing to other regions. The following is statistics of rainfall in some cities of Sichuan.

四川省部分市、州降雨量统计表
Statistics of Rainfall

市、州 Cities	成都 Chengdu	绵阳 Mianyang	广元 Guangyuan	巴中 Bazhong	南充 Nanchong	遂宁 Suining	达州 Dazhou	资阳 Ziyang	内江 Neijiang
降雨量 Rainfall (mm)	894.6	898.4	960.4	1,120.8	998.9	935.5	1,173.7	928.3	1,035.8
气候分区 Climate	湿润 Humid	湿润 Humid	湿润 Humid	潮湿 Wet	湿润 Humid	湿润 Humid	潮湿 Wet	湿润 Humid	潮湿 Wet

市、州 Cities	自贡 Zigong	宜宾 Yibin	泸州 Luzhou	乐山 Leshan	雅安 Yaan	攀枝花 Panzhihua	阿坝 Aba	甘孜 Ganzi	凉山 Liangshan
降雨量 Rainfall (mm)	1,015.8	1,110.6	1,132.7	1,295.3	1,709	1,100	775.8	808.9	1,000.5
气候分区 Climate	潮湿 Wet	潮湿 Wet	潮湿 Wet	潮湿 Wet	潮湿 Wet	潮湿 Wet	湿润 Humid	湿润 Humid	潮湿 Wet

注：降雨量>1000毫米为潮湿区，1000～500毫米为湿润区。
Rainfall>1000ml means wet，1000～500ml means humid

由上表可见，内江、自贡等市属于潮湿区，虽然资阳属于湿润区，但其降雨量已接近1000mm，仍较湿。因此，在饮食上对“防湿”“祛湿”的高度重视，成了四川地区饮食架构的重要特色。湿为阴邪，辛能发散，既可生发阳气，也能祛除湿气，配合温热之性，则辣椒、花椒等辛温的食物最有助于祛除湿邪；香能醒脾健脾，促进食欲，于是，姜、葱及桂皮等香料则成为芳香化湿之选。由此，“辛香”之好便在川菜及四川人的日常饮食中打下了深深的烙印。

From the above form, we can see that Neijing and Zigong belong to wet region. Although Ziyang belongs to humid region, its rainfall is close to 1000mm. Hence, the important characteristics of Sichuan cuisine is its emphasis on prevent dampness and dispel humidity. Humidity is yin nature. The pungent flavor can grow yang qi and dispel humidity. Food ingredients like chilly and Sichuan pepper can help to dispel humidity and evil. Fragrance can tonify spleen and promote appetizer. Food ingredients like ginger, green onion and cinnamon will remove dampness. Therefore, pungent and fragrant characteristics are the trademark of Sichuan dishes and cuisine.

三、重要节气与川菜养生

二十四节气与人体的饮食养生有着密切的联系。川菜是巴蜀地区的人们在长期的饮食实践中创造并发展起来的地方风味流派，其特点之一，就在于十分注重不同季节、不同节气对饮食烹饪的选择，以求达到养生健身的目的。下文将以春、夏、秋、冬四季为序，分别阐述在各个季节一些重要节气中的川菜养生情况。

Important Solar Terms And Regimen of Sichuan Cuisine

The 24 Solar Terms have close relations with dietary regimen of human body. Sichuan cuisine has been created on the long time dietary practice of Bashu area, which pays great attention to the options of diets in different seasons and solar terms to nourish the body. Here we'd like to introduce regimen of Sichuan cuisine in some important solar terms of four seasons, following the order of seasons.

（一）春季的重要节气与川菜养生

俗话说，“一年之计在于春”。春季养生的重要节气是立春、春分及清明。立春是春季的开始。早春时节，冬季风依然活跃，四川地形屏障的作用仍表现明显，所以气温较长江中下游同纬度地区偏高。春季在五脏中对应的是肝脏，立春开始，肝气开始生发，易出现肝气旺的情况，并影响脾胃功能，此时可适当食用“芽菜”、虾仁等食物，以助阳气生发。四川人喜食的“豌豆尖”鲜嫩可口，正是应季佳品。为了养肝护脾，春季饮食主张“省酸增甘”，南瓜、山药、粳米、谷物等甘味食物能健脾益气，效用颇佳。四川人此时爱吃春卷，春卷面皮也是甘味的代表。春卷可凉吃和油炸，易上火者则不宜选择油炸之食，四川地区早春时温度不低，脾胃强健者可凉吃。此外，川菜以味型多样著称，春季可选用甜味、咸甜、糖醋及鱼香等酸甜口味的味型。在菜品上，炒韭菜、豆芽、豌豆尖，以及椿芽烘蛋、韭黄烘蛋等都是时鲜美味。

1. Important Solar Terms in Spring and Regimen of Sichuan Cuisine

“The whole year's work depends on a good start in spring”. The important solar terms for regimen in spring are the Beginning of Spring, the Spring Equinox, Qingming. The Beginning of Spring is the start of spring. Although the winter wind is still active in early spring, the Sichuan region is still warmer than regions in the same latitude in the middle and lower reaches of the Yangtze River because the topographic barrier of Sichuan. Spring corresponds to the liver of five

internal organs. Since the Beginning of Spring, qi of liver has started to grow. It easily becomes too strong to affect the functions of spleen and stomach, hence people usually eat foods like sprout and shrimp meat to help the growth yang qi. Sichuan peoples' favorite tine pea is the seasonal food, and is tender and delicious. In order to tonify liver and protect spleen, people should "take less sour food and more sweet food". Pumpkin, Chinese yam, japonica rice and cereal are regarded sweet foods, which can tonify spleen and supplement qi. Sichuan people like to eat spring roll in this period. Crust of spring roll is also regarded as sweet food. Spring roll can be eaten fresh or fried. People with too much internal heat can not eat fried spring roll. Because it is warm in early spring in Sichuan region, people with strong spleen and stomach can eat spring roll fresh. Moreover, Sichuan cuisine is famous for varieties of tastes. In spring, people can choose sweet taste, salty sweet taste, sweet and sour taste, fish taste. As for specific dishes, Fried Leek, Fried Bean Sprout, Fried Tine Pea and Baked Egg with Chinese Toon Sprout, Baked Egg with Chive are seasonal delicacy.

春分之时昼夜平分，天地间阴阳开始趋于平衡，此时养生以平和为主，关键是注重调和阴阳，在饮食上忌大寒、大热，通过饮食搭配做到寒热互补、阴阳平衡。川菜中的各种时蔬炒菜，大多爱用姜、蒜、花椒、干辣椒炝锅，正好能综合蔬菜的寒凉之性；川人爱吃的干锅、火锅，若是搭配凉性蔬菜，也是阴阳平衡的典范。"清明时节雨纷纷"，四川更是如此，不少地区在此时节都有吃"清明果"的习俗，其中，艾叶散寒祛湿，非常适合此时食用。此时的饮食宜温不宜凉，以顾护阳气，同时还要防止肝火过旺，注意养肝清热，四川民间常吃的豆花即是清热滋阴的佳品。

The day time and night time are equal in the Spring Equinox. Yin and yang are reaching balance. Regimen in this time focuses on harmony and reconciling yin and yang. As for diet, people should avoid too cold and too hot food. People can achieve spatial balance and harmony by diet collocation. Various fried vegetables in Sichuan cuisine have used ginger, garlic, Sichuan pepper and dry chilly as spice, which can neutralize the cold nature of vegetables. Sichuan people love dry pot and hot pot, and normally eat vegetables of cold nature together to balance yin and yang. It is rainy in Qingming in Sichuan. There is the custom of eating Qingming Guo. Mugwot inside it can dispel coldness and dampness. Furthermore, people should eat warm food and avoid cold food to protect yang qi. Meanwhile, people need control the liver fire, tonify liver and clear heat. Tofu pudding is a good food to clear heat and tonify yin in Sichuan.

（二）夏季的重要节气与川菜养生

夏季尤需侧重立夏、夏至及小暑三个时节。立夏是春夏之交的标志。立夏之后要注意养护心脏、保护肠胃，注意饮食卫生。红入心，红色食物养心；夏宜"省苦增辛"，苦入心，苦味食物有助于清心去燥，但过食则会导致心气涣散、胃口不佳、损伤肺气，因此，苦味宜适量，或搭配辛味食物。炝炒苦瓜则是苦辛搭配的典范；用苦辛兼备的香料，如桂皮、陈皮等来卤、炖其他食材，是夏日川人餐桌上的美味。凉拌菜中加入姜、蒜，既能减少食材的凉性，养阳开胃，还能杀毒抑菌。桑叶芽头鲜嫩爽口，具有平肝明目、清热解毒之功，可凉拌及配搭其他食材炖煮。

2. Important Solar Terms in Summer and Regimen of Sichuan Cuisine

The Beginning of Summer, the Summer Solstice and Lesser Heat are important solar terms in summer. The Beginning of Summer is the mark of the turn of spring and summer. People should attach importance to protecting heart, protecting intestines and stomach, keeping dietetic hygiene. The red food can tonify heart. People should take more pungent foods and less bitter foods in summer. The bitter taste can help to clear the heart, but eating too much bitter food will distract the qi of heart, affect the appetizer, damage the qi of lung. Therefore, people should take proper amount of bitter food or eat pungent food together. Fried Bitter Gourd is a typical example. Sichuan people braise food ingredients with spice with bitter and pungent tastes, like cinnamon and dried tangerine peel. Adding ginger and garlic into cold dishes can not only reduce the cold nature of food ingredients, but also tonify yang, promote appetizer and kill bacteria. Sprouts of mulberry are tender and delicious, with the function of calming the liver, benefiting eyesight, clearing heat and detoxifying the body. People can braise it with other food ingredients or make it into cold dishes in sauce.

夏至时阳气最为充盛，但仍需注重保养心气。在饮食上，以清补、健脾、祛暑化湿为要点，少吃肥甘厚味之食。除红色食物外，“省苦增辛”尤重增辛，以去除渐重的湿气。此时，川人大多爱吃卤菜、冷啖杯，暗含“增辛”的养生原则。味型中带“辣”的味型也很适用，如姜汁、蒜泥、红油、麻辣等。薏仁、芡实祛湿健脾，适合湿热体质食用。到小暑时，天气开始炎热，“三伏”将至，雨水多、湿气重，饮食上需清热利湿、养心安神、健脾益气。辣椒、花椒等辛味食物虽有祛湿之功，但性温，体质燥热者需忌食或慎食。冬瓜、荷叶及淡水鱼等能利水祛湿，荷叶煮粥、蒸肉均是当令佳肴。俗话说，“小暑黄鳝赛人参”，此时鳝鱼肥美，正值食用之季。此外，应少食寒凉性食物，以免伤脾，更不宜立即食用从冰箱里取出的食物。

Although yang qi is the strongest in the Summer Solstice, people still need pay attention to protect qi of heart. Diet should focus on nourish body lightly, strengthening spleen, dispelling summer heat and humidity. People should reduce fat, sweet and thick tastes. In addition to red food, people should take more pungent foods to reduce the growing dampness. Meanwhile, Sichuan people love to eat marinated foods and cold dishes, which has followed the principle of “eating less bitter foods and more pungent foods”. Spicy taste from ginger, garlic, chilly oil is also suitable. Coix seed, gorgon fruit can dispel dampness and strengthen spleen, and are good for people with humid and heat physique. It is getting hot in Lesser Heat. With the coming of dog days, the rainfall is growing and the humidity is high. Diet should focus on clearing summer heat and promoting diuresis, nourish the heart and quiet the spirit, strengthening spleen and supplementing qi. Although pungent food ingredients like chilly and Sichuan pepper can dispel dampness, they are warm and not suitable for people with dry hot physique. Winter melon, lotus leaf and river fish can promote diuresis and dispel dampness, hence lotus leaf porridge and steamed meat are seasonable dishes. “Eel in Lesser Heat is better than ginseng”. It is the right season to eat eel, which is fat and delicious. People should control the amount of foods with cold nature to protect spleen. We should not eat foods just taken out of the refrigerator.

（三）秋季的重要节气与川菜养生

立秋和秋分是此季饮食调养的关键。立秋后天气开始转凉，但南方许多地区的气温更加酷热，所以，从立秋至秋分的这段日子又被称为“长夏”。四川地区湿度大，表现为更加闷热。脾喜燥恶湿，若脾为湿困，则表现为食欲下降、腹胀等，因此，祛湿健脾为这一时节的饮食养生要点。冬瓜、丝瓜、鲫鱼、鲤鱼以及用薏仁、赤小豆煲汤、煮粥均可；肌体阳虚者可选用花椒、辣椒、胡椒等，若辅以其他食材成菜，则能收芳香醒脾之效。此外，“长夏省甘增咸”，故应少吃精制甜食以避免困脾生湿；在出汗过多时，血压不高者可适当增加咸味调味品的用量，以帮助利水渗湿。

3. Important Solar Terms in Autumn and Regimen of Sichuan Cuisine

The Beginning of Autumn and the Autumn Equinox are the critical time of dietary regimen. It is getting cooler in the Beginning of Autumn, but it is hotter in some southern areas. Hence, days from the Beginning of Autumn to the Autumn Equinox are called “long summer”. It is sultry in Sichuan because of high humidity. Spleen like dryness and hate dampness. If it is too damp for spleen, people will suffer from loss of appetite and physogastry. The focus of dietary regimen are dispelling dampness and strengthening spleen. People can make soup or porridge with winter melon, towel gourd, crucian, coix seed and red bean. People with yang deficiency can make dishes with Sichuan pepper, chilly, pepper and other food ingredients to enliven spleen through fragrance. Moreover, “people should eat less sweet foods and more slaty foods in long summer". Eating less sweet foods can avoid the weakness of spleen and humidity. When sweating too much, people who without high blood pressure can add more salty food ingredients to promote diuresis.

秋雨绵绵是四川地区的气候特征之一，故湿度较大，与北方秋高气爽大不相同。因此，与其他地区重在防燥不同，四川地区的人们在饮食上需祛湿与润燥并重，并且要因人而异。体质燥热者可食用百合等润燥滋肺；体内有湿者仍需祛湿，为防祛湿太过而耗伤阴液，可在食用辛味时辅以酸味，川菜的酸辣味型最为适宜。此外，秋季宜"省辛增酸"，所以可多食略酸的水果，另外，四川人爱做的泡菜、泡椒和鱼香味型也很适宜。

The autumn rain goes on and on in Sichuan. Unlike the northern region, the humidity of Sichuan is still high. Hence, different from the custom of preventing dryness in other regions, Sichuan people have to dispel dampness and moisten dryness together. It varies from person to person. People with dry physique can eat lily to moisten dryness and lung. People with humid physique should dispel dampness. In order to prevent hurting yin, people can eat pungent foods with sour foods together. The sour and hot taste is suitable in this season. Moreover, "people should eat less pungent foods and more sour foods" in autumn. It is suitable to eat sour fruits, pickled chilly, foods with fish taste and pickles made by Sichuan people.

（四）冬季的重要节气与川菜养生

冬季防寒，立冬、冬至、小寒三个节气是调养重点。立冬是冬季的开始，万物收藏，需要避寒。此时，雨量渐少，但云层较厚。四川地区冬季阴冷的特点尤为突出，在饮食上可选择牛羊肉等温热性食物以温补。此外，辛味发散，可耗气伤阴，不利于阳气的潜藏及阴液的滋养，因此，花椒、辣椒等辛味食物不宜过多，烹饪方法上以蒸、煮、煨、炖为宜。冬季需"省咸增苦"，油菜、瓢儿白等蔬菜正当季，且略具苦味，暗合"增苦"的饮食养生原则。

4. Important Solar Terms in Winter and Regimen of Sichuan Cuisine

For cold protection, the Beginning of Winter, the Winter Solstice, Lesser Cold are the important solar terms. The Beginning of Winter is the start of winter. All things on earth need slumber to avoid coldness. Although the cloud is thick, rainfall is getting less. Sichuan region is typically cold and gloomy. Hence people should take warm foods like mutton and beef to tonify the body. Moreover, the pungent taste will disperse the internal heat and hurt yin and qi. It is adverse to the store of yang qi and tonify yin liquid. Therefore, people should take less pungent foods like Sichuan pepper and chilly. As for the way of cooking, steaming, boiling, stewing and braising are good in this season. "People should eat less salty foods and more bitter foods" in winter. Rape and bok choy are seasonal vegetables, which are bitter and following the above principle.

冬至是北半球白天最短、夜晚最长的一天。此时，阳气初生，需要精心保护与调养，还需养肾填精、滋养气血。四川地区在此时流行吃羊肉汤锅，既能温补阳气、健脾益胃，又能养血润燥，搭配时令蔬菜食用还能防止内热。此外，黑入肾，还可选用核桃等黑色食物。小寒标志着一年中最冷日子的开始，往往有"小寒胜大寒"之说，四川地区也是如此。寒邪伤阳，其性收敛，容易使气血凝滞。此时，人们除选用温热性食物外，还常用当归、黄芪等药食两用食材煨炖成菜，以增加温补气血的功效。

In the Winter Solstice, the day time is the shortest and night time is the longest in a year in the Northern Hemisphere. The yang qi starts to grow in this time and need to be protected and tonified carefully. People should tonify kidney, supplement essence, tonify qi and blood. Mutton pots are popular in Sichuan region in this season, which can warm yang qi, strengthen spleen and stomach, tonify blood and moisten dryness. Eating it with seasonal vegetables can prevent internal heat. Moreover, the dark food can benefit kidney. People can choose foods with dark color like walnut. The Lesser Cold marks the start of the coldest days in a year. "Lesser Cold is more important than Greater Cold" especially in Sichuan. Coldness will hurt yang and it will frozen qi and blood with the nature of convergence. In this period, people should eat foods with warm nature, and braise dishes with herbs like angelica sinensis and astragalus membranaceus which can tonify qi and blood.

四、四川节气饮食养生的特点

常言道，“天时、地利、人和”，节气可谓是中国古人顺应“天时”的经验总结，季节不同、节气不同、气候不同，食物出产也就有所不同，饮食养生故须“因时施膳”，随机应变。在四川盆地，由于其特殊的地理环境，形成了云多雾重，日照少，全年湿度较大，秋冬季节尤其阴冷的气候特点，因此，与全国其他地区不同的是，四川人在饮食上特别注重通过辛香食物以祛湿，同时用温热食物以养阳。川菜历经数千年的发展，将“天人相应”的生态观融入其中，并付诸实践，逐渐形成了许多与二十四节气密切相关的饮食养生经典菜品和习俗，是因时、因地、因人施膳的典型代表。

The Characteristics of Sichuan Dietary Regimen in Solar Terms

As the saying goes, "The time, the place and the people are in harmony." the solar terms can be described as the experience of the ancient Chinese in complying with the the nature. Solar term is the summary of ancient Chinese's experience of "follwing nature". Food ingredients are different in different seasons and solar terms. Dietary regimen should follow the principle of "taking foods according to time and seasons". In Sichuan Basin, because of its particular geographical environment, it is more cloudy, foggy, humid, and it is especially gloomy and cold in autumn and winter. Therefore, comparing to other regions in China, Sichuan diet pays more attention to dispel dampness through pungent food, and to tonify yang through warm food. After several thousand years' development, Sichuan cuisine has absorbed the ecological view of "correspondence between man and universe". There are lots of customs and dishes of dietary regimen closely connected with the 24 Solar Terms, which are typical representatives of taking foods according location, time and individual.

二十四节气与四川茶酒
The 24 Solar Terms and Sichuan Tea, Sichuan Liquor

茶和酒是中国的两大饮料。四川是中华茶文化的发源地，是中国最早种茶、饮茶、售茶的地区之一，也是中国重要的名酒产地，素有“名酒之乡”的美誉。四川人在数千年的生产劳作和繁衍生息里寻找着与自然和谐相处的方式，茶、酒也不例外。春生、夏长、秋收、冬藏，四川人选择了因时应季而饮，在二十四节气中用川茶、川酒与自然和谐共生，天人相应。

Tea and Chinese liquor are the two main beverages in China. Sichuan is the birthplace of Chinese tea culture, and one of the oldest regions where Chinese people plant tea, drink tea and trade tea. Sichuan is also the most important production place for famous Chinese liquor, and is called “Home of Famous Chinese Liquor”. Sichuan people have found the harmony with the nature during thousands of years of producing and multiplying. Tea and Chinese liquor are parts of the harmony. Following the rule of “getting born in spring, growing in summer, harvesting in autumn, storing up in winter”, Sichuan people choose the seasonal drinks and seek the harmony with the nature in the 24 Solar Terms by tea and Chinese liquor.

一、二十四节气与川茶

四川地形复杂，地势高、起伏大，气候特点是春早，夏热，秋雨，冬暖，年平均气温15℃～18℃，年降水量700～1 000毫米，水热条件优越，气温较高，阴雨天和雾天较多，适合各类茶树生长，自古以来就是全国知名的茶叶产区。除了采摘茶树叶制茶之外，人们还采用其他一些植物的茎、叶来冲泡饮用，称作“代茶饮”，它们都是川茶的重要组成部分，与二十四节气有着较为密切的关系。在此选择一部分与重要节气相关的茶品进行介绍。

The 24 Solar Terms and Sichuan Tea

Sichuan’s topography is complicated. The land is high and the landform undulates terribly. In Sichuan, spring is early, summer is hot, autumn is rainy, winter is warm. The yearly average temperature here is from 15℃ to 18℃. The yearly rainfall is from 1,000mm to 700mm. Its hydrothermal condition is good. Temperature here is comparatively high, and there are more rainy days and foggy days in Sichuan region, which are suitable for the growth of various tea trees. Since ancient time, Sichuan has been the famous tea producing area all over China. In addition to making tea with tea leaves, people also use stems and leaves of other plants, which is called “substituting tea”. All of them are important components of Sichuan tea, and have close relations with the 24 Solar Terms. Here, we’d like to choose some kinds of tea which have been connected closely with the important solar terms, and give readers related introductions.

1.清明、谷雨时节的明前茶、雨前茶

春季温度适中，雨量充沛，茶树经过冬季休养生息，在色泽、口味、香气上达到了最好状态。明朝茶人许次纾《茶疏》言，“清明谷雨，采茶之候也。”人们认为清明、谷雨时节采制的明前茶、雨前茶是一年之中茶的佳品。采之得时，饮之得时，人们便在茶的清芬甘甜中，感知节气时令的轮回。

Qingming, Grain Rain and Mingqian Tea, Yuqian Tea

The temperature is moderate and the rainfall is abundant in spring. After a whole winter's rehabilitation, tea is in the best condition in terms of color, taste and fragrance. Xu Cishu in Ming Dynasty (1368-1644) said in his *Book of Tea*, “It is the best time to pick tea in Qingming, Grain Rain”. People believe that the Mingqian Tea picked before Qingming, and the Yuqian Tea picked before Grain Rain are the best tea in a year. It is the proper time to pick and to drink. People can feel the change of solar terms in tea's clear fragrant sweet flavor.

在四川，人们最喜爱的茶叶品类是绿茶和花茶，而绿茶尤以明前茶、雨前茶为上乘。清明、谷雨时节，茶农们纷纷迎着清晨的雾气上山采茶，四川民歌唱道“采茶采茶初采茶，清明谷雨茶发芽”。此时的茶叶，叶芽稍薄，清香、鲜嫩、色碧、味透，茶叶中所含的蛋白质、多种维生素等营养成分也高于其他季节采摘的茶叶。茶叶采摘后，经过杀青、揉捻、烘焙等一系列复杂工序，便制成了清新爽口、滋味鲜醇甜润、香气醇厚的春茶。

In Sichuan, green tea and scented tea are people's favorites. Mingqian Tea and Yuqian Tea are the best of green tea. In Qingming, Grain Rain, tea farmers would go to mountains to pick up tea in morning mist. As the folk song goes, “Picking tea, picking tea, picking first tea; tea has sprouted in Qingming, and Grain Rain.” Tea leaf at this moment is fragrant, tender, tasty with thin buds. There are more vitamins inside it like protein, various vitamins than inside tea picked in other seasons. After being picked, tea need experience a series of procedures like fixation, rolling and baking, and then become spring tea with fresh sweet taste and thick fragrant smell.

明前茶、雨前茶不仅采摘、制作极为精细，饮茶也有一番讲究。谷雨这一天，除家庭饮“雨前新茶”外，还有结伴饮新茶、添乐趣的饮茶风俗，民间有“三月茶社最清出”之说。亲朋或茶友们聚集一堂，品饮清香高雅的春茶，分享品春茶的乐趣。

The way of drinking Mingqian Tea and Yuqian Tea is also sophisticated. On the day of Grain Rain, there are customs of drinking fresh tea with friends besides drinking fresh Yuqian Tea with family members. There is a folk saying that tea house would provide the best fresh tea in March. Relatives and friends will gather together to drink clear fragrant elegant spring tea, and share the great pleasure.

2.夏至、暑日时节的代茶饮

四川盆地的夏季漫长而炎热，气压低、温度高、湿度大，人体常感燥热气闷、食欲不振、肠胃不适。这个季节里，四川一些地区的人们常常在节气到来时饮用各种植物茶，以达到祛暑解热、拔毒除湿、增强身体免疫力的作用，如夏至荷叶茶、暑天三伏的“伏茶”。

Substituting Tea in the Summer Solstice and Summer Days

Summer in Sichuan Basin is hot and long. The human body will feel dry hotness, gastrointestinal discomfort and inappetence because of low atmospheric pressure, high temperature and humidity. In this season, people from some parts of Sichuan will drink various herbal teas to relieve heat, dispel humidity and boost immunity when solar terms arrive. For example, they drink Lotus Leave Tea in the Summer Solstice and “Fu Tea” Tea in dog days.

夏至时节，资阳等地的众多荷塘里莲叶田田，荷叶具有清火降压、清热养神等功效，人们将其晒干、制成“荷叶茶”，加水冲泡即可饮用。民谚说，“夏至三庚数头伏”，从小暑到处暑，四川进入了最炎热的三伏天，民间有进入大暑伏天喝“伏茶”的习俗。“伏茶”是用茶叶或祛暑化湿的中药材制成。暑湿天气喝“伏茶”，可去除暑湿邪气，避免中暑。资阳等地的人们常将苦蒿、鱼腥草、金银花一类清热凉血、解暑除湿的中草药制成茶饮用，以度过炎夏。苦蒿之味苦涩，民间流传“夏日吃苦，胜似进补”的节令养生理念，因此，苦蒿茶也成为资阳乃至川中、川南地区夏季重要的节气茶饮品种。

In the Summer Solstice, lotus bloom in lots of ponds in Ziyang region. Lotus leaves have the function of clearing internal heat, bring down blood pressure, preserving body health. People dry lotus leaves and make the Lotus Leaf Tea. As the proverb goes, "dog days will comc after the Summer Solstice". From Lesser Heat to the End of Heat, Sichuan region has entered the hottest dog days. Sichuan people have the custom of drinking "Fu Tea" in dog days. "Fu Tea" is made of tea leaves or TCM herbs which can dispel summer heat and resolve dampness. Drinking "Fu Tea" in those days can dispel summer heat and dampness, prevent sun stroke. People in Ziyang usually make tea with herbs like absinthium, houttuynia and honeysuckle which can dispel summer heat and resolve dampness. Absinthium is bitter, and people believes the theory of "eating bitter food will tonify the body in summer." Therefore, Absinthium Tea is an important substituting tea in Ziyang and central Sichuan, southern Sichuan.

3.秋分时节的柠檬茶

每年的秋分之后到霜降之前是四川柠檬大丰收的时节。安岳县盛产柠檬，其产量占全国总产量的80%以上，果实美观、品质上乘，同时也是我国唯一的柠檬生产基地县，被命名为“中国柠檬之乡”。当地人

常常将刚采摘的新鲜柠檬切片，加入蜂蜜等，用温水冲泡制成柠檬蜜茶饮用。柠檬富含维生素B_1、维生素B_2、维生素C等多种营养成分，蜂蜜富含果糖、葡萄糖和多种氨基酸、维生素。秋季饮用柠檬蜂蜜茶，有清热解毒、润燥益气等作用。

Lemon Tea in the Autumn Equinox

From the Autumn Equinox to Frost's Descent, it is the harvest time for Sichuan lemon. The production of lemon in Anyue County has taken 70% of the nation's total production.That lemon is good looking and with high quality. Anyue is the unique lemon production base county in China, and is named “Hometown of Lemon in China”. Local people will make lemon honey tea with fresh lemon slices and honey. Lemon is rich in vitamin B_1, vitamin B_2, vitamin C, etc. Honey is rich in fructose, glucose, various amino acids and vitamins. Drinking lemon honey tea in autumn can dispel internal heat, detoxify the body, moisten dryness and tonify qi.

4.大寒时节的枸杞红枣茶

枸杞红枣茶是四川地区冬季常见的养生茶饮，最适宜在大寒时节饮用。此时，天气寒冷，阴气极盛，但阳气也在逐渐生发，人们常将枸杞、红枣与冰糖同煮，有时还略加姜片，制成枸杞红枣茶来饮用，不仅可以驱寒生暖，还有养肝明目、润肺滋阴、补肾健脾以及养颜悦色等功效。

Lycium Red Date Tea in Greater Cold

Lycium Red Date Tea is a popular drink for winter in Sichuan region. It is better to drink in Greater Cold. In this cold period, yin qi is extremely strong, but yang qi begins to grow. People usually cook lycium, red date and rock candy together. Sometimes, they may add ginger slice into it. Drinking that tea can dispel coldness, tonify liver, improve eyesight, moisten lung and yin, nourish kidney and spleen, improve skin.

二、二十四节气与川酒

酒，作为粮食、瓜果提炼出的精华，原本来自土地，依赖于土地。粮食丰收，才能有美酒，正所谓美酒“天成”。这“天成之美”，正是四川以及中国人根据二十四节气的自然规律持续不断地辛勤劳作，才使得“五谷丰登”“瓜果满园”，也才有高质量的粮食和花果等食材用以酿造美酒，川酒也因此与二十四节气有了较为密切的关系。在此选择一部分与重要节气相关的酒品进行介绍。

The 24 Solar Terms and Sichuan Liquor

Chinese liquor is regarded as the essence of grains and fruits from the land. Only good harvest can produce good liquor. That is to say that the good liquor is granted by the nature. By working hard and following the 24 Solar Terms, law, Chinese people including Sichuan people can get good harvest, and can brew good liquor from grains and fruits with high quality. Therefore, Sichuan liquor has close relations with the 24 Solar Terms. Here, we’d like to choose some kinds of Sichuan liquor which have been connected closely with the important solar terms, and give readers related introductions.

1.立春时节屠苏酒

立春节气在许多年份与农历的大年三十和正月初一相近。在此时节，人们常常制作和饮用“屠苏酒”，意在益气温阳、祛风散邪。“屠苏酒”是用大茴香、荜茇、蜀椒等多种药物浸酒而成，在汉末魏晋时期已出现，但是，相传到唐代才因孙思邈而走进千家万户。孙思邈深知屠苏酒的功效，每年腊月都制作药包送给乡邻，让他们泡酒后饮用以避除疫疠之邪，随后便流传开来。因孙思邈的药屋名“屠苏屋”，此酒又常在除夕饮用，所以人们便称之为“屠苏酒”“岁酒”。饮屠苏酒的习俗与普通酒不同，不是先长后

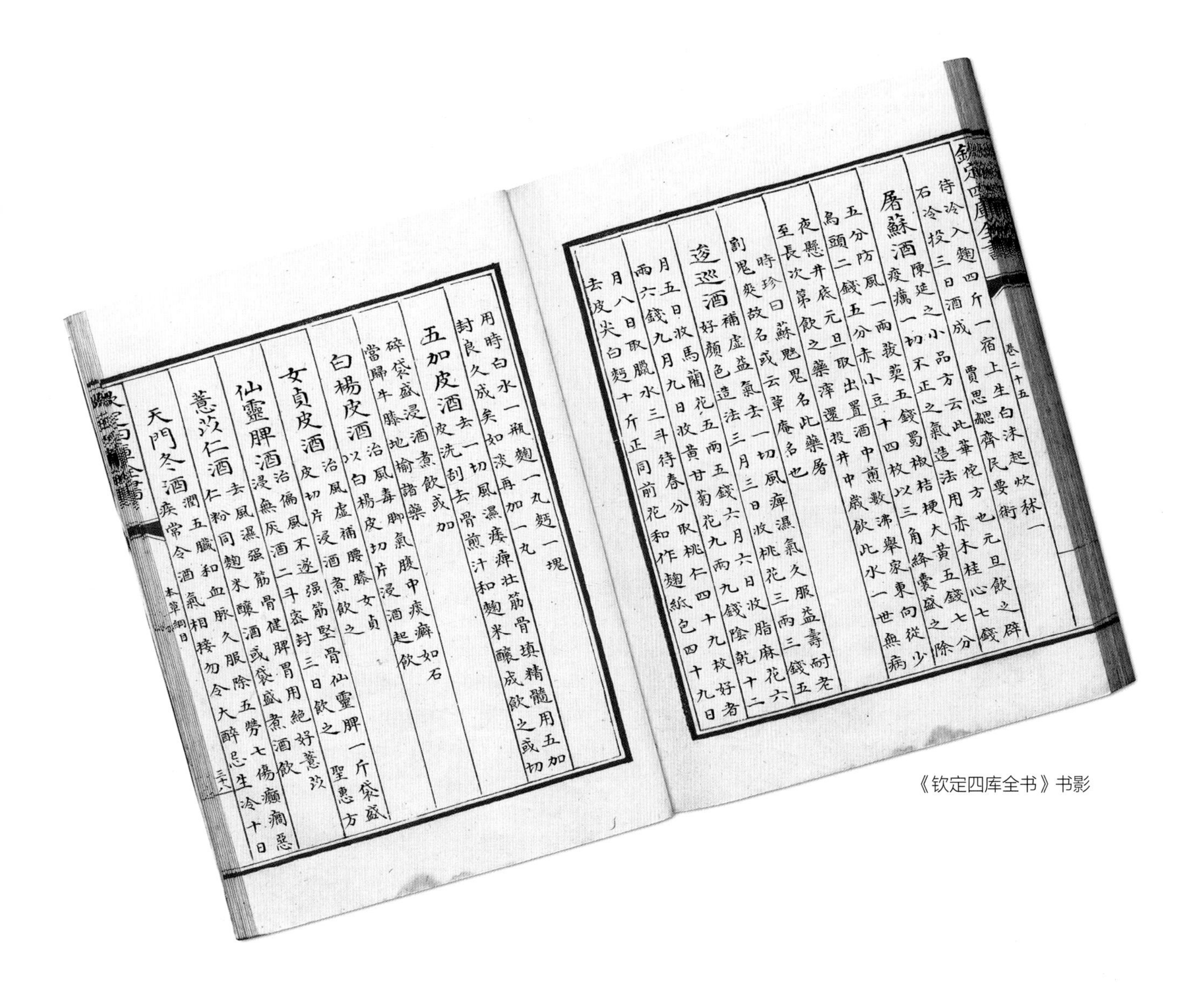
欽定四庫全書　卷二十五
待冷入麴四斤一宿上生白沫起炊秫一
石冷投三日酒成　賈思勰齊民要術
屠蘇酒　陳延之小品方云此華佗方也元旦飲之辟
疫癘一切不正之氣造法用赤木桂心七錢
五分防風一兩菝葜五錢蜀椒桔梗大黃五錢七分
烏頭二錢五分赤小豆十四枚以三角絳囊盛之除
夜懸井底元旦取出置酒中煎數沸舉家東向從少
至長次第飲之藥滓還投井中歲飲此水一世無病
時珍曰蘇魁鬼名此藥屠
割鬼爽故名或云草庵名也
逡巡酒　補虛益氣去一切風痺濕氣久服益壽耐老
好顏色造法三月三日收桃花三兩三錢五
月五日收馬蘭花五兩五錢六月六日收脂麻花六
兩六錢九月九日收黃甘菊花九兩九錢陰乾十二
月八日取臘水三斗待春分取桃仁四十九枚好者
去皮尖白麪十斤正同前花和作麴紙包四十九日

欽定四庫全書
用時白水一瓶麴一丸麪一塊
封良久成矣如淡再加一丸
五加皮酒　去一切風濕痿痺壯筋骨填精髓用五加
皮洗刮去骨煎汁和麴米釀成飲之或切
碎袋盛浸酒煮飲或加
當歸牛膝地榆諸藥
白楊皮酒　治風毒腳氣腹中痰癖如石
以白楊皮切片浸酒起飲
女貞皮酒　治風虛補腰膝女貞
皮切片浸酒煮飲之
仙靈脾酒　治偏風不遂強筋堅骨仙靈脾一斤袋盛
浸無灰酒二斗密封三日飲之　聖惠方
薏苡仁酒　去風濕強筋骨健脾胃用絕好薏苡
仁粉同麴米釀酒或袋盛煮酒飲
天門冬酒　潤五臟和血脉久服除五勞七傷癲癇惡
疾常令酒氣相接勿令大醉忌生冷十日
本草綱目
三十八

《钦定四库全书》书影

幼，而是由幼及长，因为新一年的到来，意味着孩子长大一岁，老人寿命减少一年，所以老人要最后才饮，以祝愿老人长寿。宋朝四川人苏辙《除日》言："年年最后饮屠酥，不觉年来七十余。"在四川地区，人们直到清末民国时期，依然在制作和饮用屠苏酒。1909年《游艺》第六期刊载庆余《成都月市竹枝词》之一描述道："千门腊酒饮屠苏，利市争喧闹岁除。"

The Beginning of Spring and Tusu Liquor

The Beginning of Spring is close to Chinese New Year's Eve and Lunar January First. In this solar term, people will make and drink Tusu Liquor to tonify yang and qi. Tusu Liquor is made of anise, fructus piperis longi, Sichuan pepper and liquor. It has been invented at the end of Han Dynasty and Wei and Jin Dynasties. It is said that medical scientist Sun Simiao has introduced it to common people in Tang Dynasty. Sun knew the function of Tusu Liquor well. He sent herbal packages to neighbours every December in lunar year, and ask them to soak them with liquor. Thereafter, the ways of making and drinking this liquor have spread. Sun's clinic was called Tusu House and the liquor is drunk in Chinese New Year's Eve, hence people call it Tusu Liquor or Year liquor. The habit of drinking Tusu Liquor is different from other liquor. It from the young to the elderly. The arrival of new year means that the kids are one year older and the lifetime of the elderly has lost one year. Therefore, the elderly should be the last to drink the liquor and it is a good wish for longevity. Sichuan poet Su Zhe in Song Dynasty said in his *Chu Ri*, "I am the last one to drink Tusu Liquor; now I am more than 70". In Sichuan, people kept on making and drinking Tusu Liquor till the end of Qing Dynasty and the Republic of China period. A poem published in *Entertainment* magazine in 1909 said, "thousands of families drink Tusu Liquor; the new year fair is really lively".

2.春分时节社酒

春分前后，人们迎来了祭祀土地神的日子，即“社日”。在古代农业社会，农民的收成寄托于天地神明，社神主司农事，是农事活动和农民的保护神。在万物复苏的春社中，人们祈祷社神保佑，以便有一个五谷丰登的丰收年。在社日祭祀时，供品极为丰富，包括社酒、社肉、社饭、社面、社糕、社粥等。其中，“社酒”是各家各户在头年冬季酿造封存至春社日供奉社神的酒。当祭祀活动结束后，大家则聚在一起分享供品，饮酒庆贺。

She Liquor in the Spring Equinox

Around the Spring Equinox, it is "She Ri" when people will worship god of land. In ancient agricultural society, the harvest was on the God. The She God is in charge of agriculture and is the protector of farmers. In "Chun She", people will pray for the blessing of She God and a good harvest. The offerings on "She Ri" is very abundant, including She Liquor, She Meat, She Rice, She Noodle, She Cake, She Porridge, etc. She liquor for She God is brewed and stored up in last winter. When the ceremony finishes, people will gather to share the offerings and drink the liquor together.

3.立夏时节桑葚酒

四川地区夏季的应季果实种类丰富、质优味美，桃、李、杏、枇杷、西瓜、葡萄、桑葚等琳琅满目。因为水果的营养价值高、甜美爽口，人们常在家中将各类水果自酿为果酒饮用。在立夏前后的资阳地区，尤其是乐至县桑树种植量大，此时桑葚成熟，许多人常采桑葚酿造桑葚酒，享受自然的馈赠。桑葚是桑树的果实，富含蛋白质、多酚类黄酮、花色苷及多种维生素和微量元素，易被人体吸收，具有很高的营养价值，被称为“民间圣果”。用桑葚酿造的桑葚酒果香浓郁、口感清爽甜美，具有提高人体免疫力、强身抗老、美容养颜等作用。

Mulberry Wine in the Beginning of Summer

In summer, there are lots of seasonable fruits in Sichuan region, like peach, plum, apricot, loquat, watermelon, grape and mulberry. People usually brew fruit wine at home with those sweet nutritious fruits. In Ziyang region around the Beginning of Summer, there are lots of mulberry trees in harvest time. People will pick mulberry and brew mulberry wine. Mulberry is rich in protein, polyphenolic flavonoids, anthocyanin. It is digestible and nutritious. Mulberry wine is thick, fresh and sweet, which can boost immunity, strengthen health, resist aging and maintain beauty.

4.寒露时节菊花酒

寒露前后，在许多年份常常迎来农历九月九日的重阳节。此时已是深秋，草木黄落、百花凋谢，唯有菊花绽放，人们在登高、赏菊之时会选择饮用菊花酒助兴。汉朝刘歆《西京杂记》言：“菊花舒时，并采茎叶，杂黍米酿之，至来年九月九日始熟就饮焉，故谓之菊花酒。”意思是在菊花盛开之时，将菊花连枝带叶采下来，与黍米放在一起酿造，等到下一年九月九日便成熟了，即可饮用。如此酿造菊花酒费时较长，还有一种简便方法，即是将菊花的花瓣直接放入酒中即成。四川地区的人们也常常在重阳节酿制、饮用菊花酒。清朝四川梁平人涂宁舒《竹枝词》言：“无端风雨满江皋，黄菊花开引兴豪。烧酒酿成蔬菜熟，相携明日好登高。”深秋中登高望远、畅饮菊花酒，既享受秋日景色，也祈求健康长寿。

Chrysanthemum Wine in Cold Dew

Around Cold Dew, it is Double Ninth Festival in many years. It is late autumn, plants have turned yellow, and all flowers fade except for chrysanthemum. People will choose to drink chrysanthemum wine during climbing mountains and appreciating chrysanthemum. Liu Xin in Han Dynasty has said in his *Records of Western Capital*, "when chrysanthemum blossoms, people pick its flower, stem and leaf, and then add millet to brew wine. It can be drunk on next 9th September. It is called chrysanthemum wine". There is a simpler way, just adding the petals into the liquor. Sichuan people also make and drink chrysanthemum wine in Double Ninth Festival. Tu Ningshu from Liangping in Qing Dynasty has described this scene in his *Bamboo Branches Poem*. In late autumn, people will climb mountains to enjoy autumnal landscape, and will drink chrysanthemum wine freely to pray for health and longevity.

5.立冬时节启冬酿

立冬之日开始酿酒，是中国许多地区传统的酿酒风俗。冬季水体清冽，气温低，可有效抑制杂菌繁育，又能使酒在低温长时间发酵过程中形成良好风味，是最适合酿酒的季节之一。中国古代许多地区自立冬之日开始酿酒，至立春后则酿成、饮用，人们常常称之为“春酒”。《诗经·国风·七月》言：“十月获稻，为此春酒，以介眉寿。”因这种酒是冬季酿造，又称为“冬酿酒”，可以用稻谷，也可以加草药酿制。清朝顾禄《清嘉录》载“乡田人家，以草药酿酒，谓之冬酿酒”“十月造者，名‘十月白’”。在四川资阳等地，许多人在立冬之后则喜爱用当年秋季丰收的糯米加酒曲酿制口味香甜醇美、酒精度低的米酒，当地称为“醪糟”，在冬季享用极佳。

Starting to Brew Liquor in the Beginning of Winter

Starting to brew liquor on the day of the Beginning of Winter is a tradition in many regions in China. Water in winter is clear and cold, which can control bacteria growth and form good flavor. Hence winter is one of the best seasons to brew liquor. In ancient China, people brew liquor on the day of the Beginning of Winter and drink the liquor after the Beginning of Spring, which is called Spring Liquor. The *Book of Songs · National Customs · July* said, "harvest grain in October and brew Spring Liquor with it." It is also called Winter-Brewing Liquor because it is brewed in winter. Furthermore, people may add herbs into the liquor. Gu Lu in Qing Dynasty has said in his *Qing Jia Record*, "rural families brew the liquor with herbs", "the liquor brewed in October is called October White". In Ziyang, many people like to drink rice liquor after the Beginning of Winter. That liquor is made of newly harvested glutinous rice and distiller's yeast, and local people call it "Lao Zao" in Chinese.

总之，从立春到大寒，二十四节气循环不殆，草木有枯荣，身体有盛衰，生命却在不断追求。茶与酒作为四川人生活中最重要的饮品，更是"顺天应时"之一文化精神的完美体现，茶酒生活展现出了四川人遵循自然规律、尊重生命节奏的生存智慧。一节一气，一茶一酒，四川人在氤氲的茶香与醇美的酒香中感应天地万物，四季轮回，与自然和谐相处，温暖共存。

In a word, the 24 Solar Terms keep on cycling from the Beginning of Spring to Greater Cold. Although there are ups and downs for both plants and humans, life is developing all the time. Tea and Chinese liquor are the most important beverages for Sichuan people, which have reflected Sichuan people's wisdom of following natural laws and respecting life. One tea and one liquor are for one solar term. Sichuan people are feeling all the living things, seasonal cycling and harmony with the nature in the fragrance of tea and Chinese liquor.

伍

二十四节气与四川食俗

The 24 Solar Terms and Sichuan Eating Customs

俗语道，“冬至馄饨，夏至面”。二十四节气，不仅是中国人的一种时间法则，也是我国传统阴阳合历历法制度（俗称“农历”）的重要组成部分，既与农业生产密切相关，同时还具有丰富的民俗内涵，早已深入中国人的社会生活，影响着中国人的思维和行为方式，因此，一些民俗学家将“二十四节气”称为“民俗系统”。同时，二十四节气与中国传统节日也存在着不同程度的关联，尤其是在饮食民俗上的表现更为明显。

As the proverb goes, “eating wonton in the Winter Solstice and eating noodle in the Summer Solstice”. The 24 Solar Terms are the law of time and the important component of our lunar calendar, which are closely connected with agricultural production. They have abundant folklore connatations, which have deeply influenced Chinese society and Chinese way of thinking and behavior. Hence, some experts on folklore have called the 24 Solar Terms the folklore system. Meanwhile, the 24 Solar Terms have different connections with Chinese traditional festivals, which have more obvious reflections on eating customs.

一、二十四节气、传统节日与饮食民俗的关系

节气与中国传统节日经历了一个由合到分的演变。在远古时代，人们观象授时，农事周期就是庆典周期，节气最初也就是节日，二者相融。我国的历法，自秦汉以前的古六历开始，到汉代的太初历，再到清末天历止，共有百余种历法，但都是阴阳合历。其中，专家学者认为，人们的庆典仪式依据阴历而固定周期，节气则按照阳历固定周期，于是在历史长河中节气与节日便逐渐发生了一定的分离，不过，总体上依然保持着不同程度的关联。其中，一些节气长期受到高度重视，有隆重或大范围的庆典及习俗，因此一直作为节日保留传承至今，如“冬至”就是节气与节日长期合一的典范，它自古以来在全国范围内都受到高度重视并有相关庆典及习俗；其次是“清明”，在唐朝及其后的时间里不断受到重视，民俗活动突出，也可以说是节气与节日合一，最终发展成为中国四大传统节日之一。而另一些节气因庆典或习俗活动逐渐弱化，甚至被迁移到其他节日举行，使得这一节气与节日有了较大分离，因而仅作为节气保留传承下来，如“秋分”就是节气与节日分离的典型，它在唐宋以前是举行祭月的重要之日，但在唐宋以后因祭月之礼逐渐迁移到中秋，便主要作为节气传承至今。

The Relations among the 24 Solar Terms, Traditional Festivals and Eating Customs

Solar terms and Chinese traditional festivals have experienced the evolution from unification to separation. In ancient times, people fixed the time by observing celestial phenomenon. Agricultural cycling is the celebration cycling, and the solar term is festival too. China has more than 100 calendars in history, from Gu Liu Calendar in Qin and Han Dynasties to Taichu Calendar in Han Dynasty, and to Tian Li at the end of Qing Dynasty. All of them are mixture of lunar calendar and solar calendar. Experts believe that people fix the cycle of ceremonies according to lunar calendar and fix the cycle of solar terms according to solar calendar. Then solar terms and festivals gradually separate in the long history, but there are still some kinds of connections. Some terms have got importance for a long time with great ceremonies and customs, therefore they are preserved as festivals till now. The Winter Solstice is a good example. It has been emphasized on with related ceremony and custom since ancient time. Qingming is another example. It has received attention since Tang Dynasty and evolved into one of the four main Chinese traditional festivals. Some solar terms' ceremonies and customs have been shifted into other festivals, then they have separated from each other. The Autumn Equinox is a good example. Before Tang and Song Dynasties, it is an important time to worship the moon. After Song Dynasty, the ceremony has been held in Mid-Autumn Festival, and the Autumn Equinox has become a pure solar term till now.

二十四节气大多有丰富的民俗活动，大致包括有信仰与仪式民俗、饮食与养生保健民俗、观赏与娱乐民俗等，具有奉祀神灵、崇宗敬祖、除凶祛恶、休闲娱乐等功能，可达到以应天时、维护亲情、以求平安、放松心情的作用。中国人崇尚“民以食为天”，在每个节气都遵循着“天人合一”“顺应四时”的理念，并且与中国传统养生思想相结合，形成了以食材生产和饮食品制作与享用为中心、具有一定的地域特色的中国节气饮食民俗。围绕着二十四节气饮食民俗，还产生了一些民间故事传说、文人诗词歌赋等，不仅表达了人们的思想和情感，还寄托着人们的希望和追求，逐渐形成了多姿多彩的节气饮食民俗文化。

There are abundant folk activities for most 24 Solar Terms, including folk belief, folk regimen, folk amusement, which have the function of worshiping gods and ancestors, dispelling evil, relaxing. Through those folk activities, people can make harmony with the nature, maintain the family, pray for safeness, and relax the mood. For Chinese people, “food is the first necessity of the people”. In every solar term, Chinese people follow the ideas of “harmony between man and nature” and “changing according to seasons”. Mixed with Chinese traditional regimen, people have formed various local solar term eating customs, which put food ingredients producing and dishes making and enjoyment to the center. Around the 24 Solar Terms, there are many stories, legends and poems, which not only express people's thinking and emotion, but also hold people's wishes and pursuit. Colorful solar term eating customs have come into being gradually.

二、重要节气与四川饮食民俗

在二十四节气的相关民俗中，饮食民俗是最受人喜爱、较易于传承的一类。它是中国各地人们节气饮食生活的经验和智慧，遵循着传统时令饮食原则“必先岁气，毋伐天和”。这里以春夏秋冬四季为序，分别阐述各个季节中一些重要节气有关四川的饮食习俗。

Important Solar Terms and Sichuan Eating Customs

Among folklore related to the 24 Solar Terms, eating custom is the most popular and easy to carry forward. It is the wisdom and experience of eating activities in solar term, following the traditional principle of “understanding the law of seasonal change before taking actions and going with the flow”. Here, we will follow the order of seasons, and introduce Sichuan eating customs related with important solar terms in every season.

1.春季重要节气与饮食民俗

“一年之计在于春”，春季大地回暖，阳气逐渐上升，春季节气的饮食民俗主要围绕着助阳迎春展开。其中，重要节气有立春、春分、清明等，都有不同的饮食民俗。

月令輯要 卷五 十

皆賜春幡勝以羅爲之宰執親王近臣皆賜金銀幡勝[illegible]雲圖乍翻紅菜甲綵幡新剪綠楊絲[illegible]曉開魚鑰朝衣集綵勝飄揚百辟冠[illegible]春幡春勝一陣春風吹酒醒[胡[illegible]]最好是戴綵幡春勝釵頭雙結

食雞辟疫 增[肘後方]冬至日取赤雄雞作腊至立春日煮食至盡勿分他人辟禳瘟疫

迎春髻 增[炙轂子錄]漢之迎春髻立春日戴[元好問詩]金釵翠鬟迎春髻銀燭光摇午夜花

氣首 增[五代史司天考]晉司天監馬重績起唐天寶十四載乙未爲上元用正月雨水爲氣首

春帖子 增[溫公日錄]翰林書待詔請春詞以立春日剪帖於禁中門帳皇帝閤[乾淳歲時記]學士院撰進春帖子帝后貴妃夫人諸閣各有定式絳羅金縷華粲可觀

春盤 原[四時寶鑑]東晉李鄂立春日以蘆菔芹芽爲菜盤相饋貺[又]立春日薦春餅生菜號春盤增[乾淳歲時記]後苑辦造春盤供進及分賜貴邸宰臣巨璫翠縷紅絲金雞玉燕備極精巧每盤直萬錢

御製春盤詩小摘園中露新香糝作羹歲和欣有兆甘菜巳春生 增[歐陽詹春盤賦]假盤盂而作地疏綺繡以爲珍叢林具秀百卉爭新[杜甫立春詩]春日春盤細生菜忽憶兩京梅發時盤出高門行白玉菜傳纖手送青絲[蘇軾詩]青蒿黃韭簇春盤[陸游立春詞]春盤春酒年年好試戴銀旛判醉倒

月令輯要 卷五 正月令 節序 七

《月令辑要》清康熙内府四色套印本

Important Solar Terms and Eating Customs in Spring

"Spring is the beginning of the whole year", the land is getting warm and yang qi gradually grows in spring. Eating customs in spring are mainly for helping yang and welcoming spring. The Beginning of Spring, the Spring Equinox, Qingming are important solar terms and have different eating customs.

周朝时，周天子十分重视各个季节之首，每一季节的首个节气都要举行迎接仪式。据《礼记·月令》记载，立春之日为春季之首，周天子亲率三公、九卿、诸侯大夫到东郊迎春。在四川，立春节气的迎春风俗演化出迎春、点春、春台、打春等程序，其目的在于促进食材的生产。立春的饮食也体现出迎春、助阳的性质，咬春、尝新是其重要的饮食活动。立春有春盘，也叫“五辛盘”，人们以五种辛辣之物发五脏之气，如葱、姜、蒜、韭菜、萝卜等，“取迎新之义”。立春还有春卷，萝卜也是立春的应节食品，古时人们无论贵贱都嚼萝卜，称为“咬春”，以除却春困。春分是春季的又一个重要节气。春分节气前后与“春社”相邻，四川地区在古代要祭祀句芒神，有“敬春分馍馍”之俗。在这一天，农民不到田间劳作，而且将糯米粉做成团、加入艾蒿制成馍馍，祭祀雀鸟，希望它们不食农作物。许多四川农家在这一天也吃炒豆，期望减少农作物的虫害，还吃菜卷子、河鲜等。此外，四川地区历来非常重视水利，清明时节，一年一度的“放水节”是食材生产的重要习俗之一。清明节这天，人们除了祭祀祖先，大多还带着酒肴出城游玩聚餐，举办“清明会”，吃清明糕、馓子等美食。

《农书》元·王祯著　明嘉靖九年山东布政司刊本

In Zhou Dynasty, the King attached great importance to the beginning of every season. There was welcome ceremony in the first solar term of every season. According to *The Book of Rites · Yue Ling*, the day of the Beginning of Spring is the first day of spring, and the King will carry his courtiers to welcome the spring in eastern suburbs. In Sichuan, the customs of welcoming spring in the Beginning of Spring have evolved into Ying Chun, Dian Chun, Chun Tai and Da Chun. The purpose is to promote the food ingredients producing. Foods in the Beginning of Spring have the nature of welcoming spring and helping yang. Yao Chun and taste a fresh delicacy are the important eating customs. There is Spring Plate in the Beginning of Spring, which is called Wu Xin Pan too. People believe we can help the qi from the five internal organs to grow by eating five pungent foods, like green onion, ginger, garlic, leek, radish. It also means "welcoming new arrivals". Spring roll and radish are seasonal foods in the Beginning of Spring. In ancient times, people no matter rich or poor bit radish to drive away spring drowsiness, which is called Yao Chun. The Spring Equinox is another important solar term in spring. The Spring Equinox is close to Chun She. Ancient Sichuan people would worship Goumang God. There was a custom of offering the Spring Equinox steamed buns. On that day, farmers stop working and make steamed buns with glutinous rice flour and mugwort. Those steamed buns will be offered to birds. Farmers hope the birds would not eat crops. In addition to vegetable rolls and freshwater fishes, many farmers hope to control the pest by eating fried beans on that day. Moreover, Sichuan people always pay great attention to water conservancy. The yearly Water Festival is an important custom for producing food ingredients in Qingming. On the day of Qingming, after worshiping ancestors, people will play outside the city carrying dishes and liquors. They will hold Qingming party, eat Qingming cake, and taste other foods like San Zi.

2.夏季重要节气与饮食民俗

夏季是阳气高涨的时节，迎夏与度夏是夏季节气饮食民俗的主要内容。其中重要的节气是立夏、夏至等。

Important Solar Terms and Eating Customs in Summer

Yang qi has grown rapidly in summer. Welcoming summer and spending summer are the main contents of eating customs. The Beginning of Summer and the Summer Solstice are the important solar terms in this season.

立夏，作为夏季的开始，自古受到人们的重视，形成了许多礼仪习俗。据《礼记·月令》记载，立夏之日，周天子亲率三公、九卿、大夫到南郊迎夏。在四川很多地区，立夏时节的农家开始在田间忙碌着插秧，农谚有"立夏小满正栽秧"之语。第一天栽秧称"开秧门"，主家要先敬秧苗土地，还要备丰盛酒食招待栽秧师，并请亲友邻居来家吃"栽秧酒"，有"栽秧子是待女婿，打谷子是待舅子"的俗语。除一日三餐外，还有类似"打尖"的幺台酒。幺台一般有腌腊肉、盐蛋、煎炒菜肴和酒水，稻田离家近的在家中吃幺台，稻田离家远的则送到田间享用，以免误工。此外，四川人立夏后还喜欢用黄鳝制成美食后享用，有益于身体健康。夏至是夏季的又一重要节气，也是阴气上升的时节，主张顺气的古人在这一天要举行相应的扶阴助气仪式。从周朝起直到清朝一直保持着在夏至日举行祭祀地神的仪式，以驱除疾疫、荒年与饥饿。粽子是南北朝时期的夏至美食，后来改到了端午节。夏至极热，四川民间除了食用粽子之外，还常常食用凉面。诗圣杜甫曾作诗歌咏他在四川吃到过的"槐叶冷淘"，称赞它"经齿冷于雪，劝人投此珠"。

The Beginning of Summer has got people's attention since ancient times. There are lots of customs for the Beginning of Summer. According to *The Book of Rites · Yue Ling*, King of Zhou would lead his courtiers to welcome summer in southern suburbs. In many areas of Sichuan, farmers are busy with transplanting rice seedlings in fields in the Beginning of Summer. As the farmer's proverb goes, "it is the right time to transplant rice seedlings in the Beginning of Summer and Lesser Fullness of Grain. Transplanting rice seedlings on the first day is called "Kai Yang Men". The hosts

《耕织图》（局部）

should worship rice seedling and land first, then provide abundant foods and drinks to the labors, and invite relatives and neighbours for "Zai Yang Liquor". As the farmer's proverb goes, "receive son in law during transplanting rice seedlings, and receive brother in law during harvesting grains." In addition to main meals, there are some snacks called Yao Tai Liquor. Normally, Yao Tai is composed of Chinese bacon, salty egg, fried dishes and liquor. Some people choose to eat Yao Tai at home if the fields is close; some people choose to eat Yao Tai in the fields to save time if it is far away from home. Moreover, Sichuan people like to eat dishes made with eel after the Beginning of Summer, which is good for health. The Summer Solstice is another important solar term in summer. Yin qi has grown in this solar term. Ancient people who insisted on following qi would hold a ceremony to help yin qin growing on that day. From Zhou Dynasty to Qing Dynasty, there is a traditional ceremony to worship god of land. The purpose is to dispel diseases and famine. In the Northern and Southern Dynasties(420-589), Zongzi was a typical food in the Summer Solstice. At present, it is the food for Dragon Boat Festival. It is extremlely hot in the Summer Solstice. Sichuan people will eat cold noodles with sauces in addition to Zongzi. Du Fu, Saint of Poetry, has written a poem for a Sichuan cold noodles with locust tree leaves juice named "Huai Ye Leng Tao". He said, "it is as cold as snow, and people should share the wonderful pearl-like food".

3.秋季重要节气与饮食民俗

秋季是阴气生长的季节，秋风起，天转凉，收敛与对阴性世界的顺从是秋季节气饮食民俗表达的主要内容。其中重要的节气是立秋、秋分等。

Important Solar Terms and Eating Customs in Autumn

Autumn is the season for the growth of yin qi. When autumn wind blows, the weather is getting cold. Convergence and following yin are the main contents of eating customs in autumn solar terms. The Beginning of Autumn and the Autumn Equinox are important solar terms in this season.

《礼记·月令》载，立秋之日，周天子亲率三公、九卿、大夫到京城西郊迎秋。这一时节，农人开始收获新稻谷进献给天子，天子在尝新之前，先供奉祖先。立秋有"咬秋"民俗，人们在立秋时节吃秋瓜、秋桃，以养生避疫。这也是《诗经》所说"七月食瓜"的遗意。此外，四川也与其他地区一样，立秋还有贴秋膘、尝秋鲜之俗，以补充劳动所消耗的精力。秋分是秋季又一个重要节气，在上古时也是祭月的重要日子，朝日夕月，说的就是春分祭日、秋分祭月。此后，历代都有祭月之礼，只是唐宋以后，民间祭月大多在中秋进行，并且食用月饼、糍粑、瓜果，饮桂花酒等，其乐融融。秋分时节在许多年份又与"秋社"相邻，它是农家祈年的日子，俗称"土地诞"。在四川地区，乡间多演傩戏、陈杂供，其规模大于"春社"，也要举行宴饮活动。

The Book of Rites · Yue Ling recorded that King of Zhou would lead his courtiers to welcome autumn in the western suburbs on the day of the Beginning of Autumn. Farmers would present newly harvested grains to the King. Before tasting the grains, the King would offer them to ancestors first. There is a custom of biting autumn in the Beginning of Autumn. People will eat towel melon and peach to avoid plague. Hence, the *Book of Songs* said, "eating melon in July". Moreover, to replenish the energy used by working, there are customs of sticking autumn fat and tasting the fresh of autumn. The Autumn Equinox is another important solar term. It is the time of worshiping moon in ancient times. "Zhao Ri Xi Yue" means worshiping sun in the Spring Equinox and worshiping moon in the Autumn Equinox. After Tang Dynasty and Song Dynasty, worshiping moon has been held in Mid-Autumn Festival. People would eat mooncake, Ciba, fruits and sweet-scented osmanthus wine. In many years, the Autumn Equinox is the time to pray for good harvest and is also called "Birth of Land". In Sichuan, there are Nuo opera and offerings. People will hold banquet bigger than Chun She.

4.冬季重要节气与饮食民俗

冬季是冬藏的时节，严寒逐渐到来。人们为了顺利地度过寒冷的冬季，冬季节气饮食民俗主要围绕着扶阳助阳展开。其中重要的节气是立冬、冬至等。

Important Solar Terms and Eating Customs in Winter

Winter is the season for store. It is getting severely colder. In order to pass through the cold winter, the main contents of eating customs in winter are helping yang and supporting yang. The Beginning of Winter and the Winter Solstice are the main solar terms in this season.

据《礼记・月令》载，立冬之日，周天子亲率三公、九卿、大夫到北郊迎冬。后代帝王沿袭了立冬北郊迎气的习俗。立冬意味着冬季开始，人们要做越冬的准备。但是，从物候上看，四川许多地区从立冬至小雪之前天气仍较温暖，一些果树仍会开花，有“小阳春”之说。立冬时节在许多年份与农历十月初一日的“牛王诞辰”（牛王节）相邻，清末民国时期的四川城乡(以农村为最)皆要用糍粑祭祀牛神，以酬其辛劳。此外，在清末民国时期的资阳、内江等地，农历十月上旬，大小糖房要挑选吉日动工，以甘蔗榨糖。糖房老板要在开工这一天请工人、邻居、租牛户及青山户等喝酒吃饭，名为“起搞酒”。有的糖坊不仅产糖，还加工制作蜜饯等，使得资阳、内江拥有了最好的一种土特产品，人称“半日驱车资内过，齿牙尝遍是甜乡”。冬至是冬季的又一重要节气，是阴气高涨、阳气发生和萌生希望的时节。作为大如年的节气与节日，冬至特点之一在于要有祝贺活动，又称“拜冬”“贺冬”。四川各地祭祖最重冬至。四川地区在冬至时祭祀祖先，一般是合族聚会祠堂、祭祀祖先，然后宴饮。庆贺冬至的饮食品种有顺阳助长之义，俗语言“冬至馄饨，夏至面”。但是，在四川地区，冬至这一天还常常杀猪、做腊肉和香肠，以备来年宴客、饷农之需，民间称之为“冬至肉”。如今，四川地区的人们在冬至时盛行吃羊肉。“羊”谐音“阳”，人们认为冬至是一年中最寒冷时节的开始，吃羊肉不仅可以暖和身体，还能资助阳气生长。

According to *The Book of Rites・Yue Ling*, King of Zhou would lead his courtiers to welcome winter in northern suburbs. The Beginning of Winter means the start of winter and people should make preparations for winter. However, from the perspective of phenology, it is still warm from the Beginning of Winter to Lesser Snow in many parts of Sichuan. There is a balmy weather and some fruit trees will bloom. The Beginning of Winter is close to Ox King Festival on 1st October in lunar year. At the end of Qing Dynasty and in the Republic of China period, Sichuan people mostly in rurual areas would offer Ciba to God of OX. In that period, sugar factories in Ziyang and Neijiang would choose an auspicious time to start making sugar with sugarcane in the first ten days of October. The boss would invite workers, neighbours, ox owners for foods and drinks, which is called “Qi Gao Jiu”. Some factories produce not only sugar but also preserved fruits which are the best local speciality in Ziyang and Neijiang. People said that you can get the taste of sugar hometown after driving into Ziyang and Neijiang for half day. The Winter Solstice is another important solar term. Yin qi is extremely strong and yang qi starts to grow in this period. The Winter Solstices is as important as the new year. There are celebrations called “Bai Dong” and “He Dong” in this solar term. It is the most important time to worship ancestors in Sichuan. Relatives will gather together to worship ancestors in the ancestral temple, and then have banquet together. Foods celebrating on the Winter Solstice have the meaning of helping yang to grow. As the proverb goes, “eating wonton in the Winter Solstice, and eating noodles in the Summer Solstice”. In Sichuan, people will slaughter pigs to make Chinese bacon and sausages on the day of the Winter Solstice. Those foods are made for the coming year. Moreover, Sichuan people will eat mutton in the Winter Solstice. Mutton has the same pronunciation as yang in Mandarin. People believe that the Winter Solstice is the start of the coldest time in a year, and eating mutton can warm the body and help yang qi to grow.

三、四川节气饮食民俗的特点

四川地区二十四节气的饮食民俗和全国其他地区相比，既有相同之处，又有所差异。其相同之处在于，大都遵循了“天人合一”“顺应四时”的传统理念，体现了“春生、夏长、秋收、冬藏”的天道。不同之处在于，由于四川特殊的地理环境、农业生产等情况，在四川的节气饮食民俗中，有一部分民俗主要是围绕着水稻的生产而展开的。由于气候条件的不同，四川与同属水稻生产区的江浙、广东一带相比，也有自己的特点。以有关插秧的民谚为例，江苏有“立夏浸秧”之语，广东有“清明谷雨时，插田莫迟疑”，而四川则是“立夏小满正栽秧”。不同地区，插秧的节气与时间不同。此外，四川节气的饮食品种大多是以稻米为主制成，如春卷、清明粑粑、艾蒿馍馍、糍粑等，与北方地区大多用麦面制作节气食品有所不同，这也充分体现了中国节气饮食习俗的地域性。

Characteristics of Sichuan Eating Customs in Solar Terms.

Comparing to other regions of China, Sichuan eating customs in the 24 Solor Terms have something in common and something in different. All of them follow the ideas of “harmony between man and nature” and “changing according to seasons”, and follow the rule of “getting born in spring, growing in summer, harvesting in autumn, storing up in winter”. Because of the special geographical conditions and agricultural production in Sichuan, some eating customs are around the producing of rice. Comparing to the rice growing in Jiangsu, Zhejiang and Guangdong provinces, Sichuan rice has its own characteristic because of the particular climate. Take the case of the farmer's proverbs on transplanting rice seedlings, there is “soaking rice seedlings in the Beginning of Summer” in Jiangsu and “transplanting rice seedlings in Grain Rain and Qingming” in Guangdong. In Sichuan, it is “transplanting rice seedlings in the Beginning of Summer and Lesser Fullness of Grain”. The solar terms and time for transplanting rice seedlings are different in different regions. Furthermore, Sichuan solar term foods are mainly made with rice, like spring roll, Qingming cake, mugwort steamed bun, Ciba, etc. The northern part of China normally use flour. That is the regional phenomenon for Chinese solar term eating customs.

总之，二十四节气是中国传统社会里的一个个重要时间节点，围绕这些节点形成了一系列多姿多彩的饮食民俗活动。人们通过饮食民俗的丰富表达，力图实现与自然、与社会、与人际的和谐共融，并且期盼五谷丰登和健康幸福。

In a word, the 24 Solar Terms are important points of time in traditional Chinese society. Around those points, there are a series of colorful eating customs. People try their best to get harmony with the nature, society and among people by eating customs, and to pray for good harvest, health and happiness.

PART 2

下篇 二十四节气与川菜创意菜点

THE 24 SOLAR TERMS AND CREATIVE SICHUAN DISHES

THE BEGINNING OF SPRING

立春

香艳横枝 万木生芽

立春

唐·杜甫

春日春盘细生菜，
忽忆两京梅发时。
盘出高门行白玉，
菜传纤手送青丝。
巫峡寒江那对眼，
杜陵远客不胜悲。
此身未知归定处，
呼儿觅纸一题诗。

立春物候 | PHENOLOGY IN THE BEGINNING OF SPRING

立春是二十四节气中的第一个节气，也是春季的第一个节气。当太阳到达黄经315° 时为立春，时间为每年的2月3日～5日。在我国民间，人们一直都把立春作为孟春时节的起始，经千年而不变。元朝吴澄《月令七十二候集解》言“立春，正月节。立，建始也”“于此而春木之气始至，故谓之立也”。意思是说此时春天到来，草木开始萌芽生长，故名立春。

The Beginning of Spring is the first of the 24 Solar Terms, and the first solar term in spring. It begins when the Sun reaches the celestial longitude of 315°. It usually begins from 3rd to 5th February every year. In Chinese folk tradition, it has been regarded as the beginning of Spring for thousands of years. In his book *A Collective Interpretation of the Seventy-Two Phenological Terms*, Wu Cheng (Yuan Dynasty 1271-1368) said that spring wind has arrived and the plants have started to grow in this period, hence we call this period the Beginning of Spring.

我国古代将立春后的十五天又分为三候：一候东风解冻；二候蛰虫始振；三候鱼陟负冰。意味着立春五日后，东风送暖，大地开始解冻；再过五日，蛰居土中冬眠的虫子慢慢苏醒；再经五日，河冰渐渐消融，鱼儿开始上浮到水面游动，此时，水面上的余冰，如同被鱼儿托举一般浮在水面。立春相应的花信为：一候迎春；二候樱桃；三候望春。

Ancient China has divided the 15 days after the Beginning of Spring into 3 pentads: in the first pentad, the eastern wind clears the cold; in the second pentad, dormant insects begin to awake; in the third pentad, fishes start to swim in the river with melting ice. In other words, the warm eastern wind unfroze the ground 5 days after the Beginning of Spring; insects and worms hibernating under the ground revive slowly 5 more days later; fishes will swim in the river with melting ice and they look like pushing up the rest ice to the river's surface. The news of flowers blooming in the Beginning of Spring are as follows: winter jasmine for the first pentad, cherry for the second pentad and flos magnoliae for the third pentad.

食材生产 | PRODUCING FOOD INGREDIENTS

立春后气温回升，各种食材长势加快，人们常常抓紧时间灌溉，中耕松土，追施返青肥，以促进其生长，同时做好防寒、防冻、防虫害和冬小麦的除草等工作，防止“倒春寒”天气对食材生长造成的危害。在四川地区，人们更是抓紧时间耕翻早稻秧田，做好选种、晒种及夏收作物的田间管理。俗语说，“立春雨水到，早起晚睡觉”。立春时节，全面开展春耕春种，尤其要密切关注天气变化，及时播种早稻。

With the rise in temperature after the Beginning of Spring, crops' growth is being accelerated. People usually seize the time irrigating, loosening the soil, fertilizing in turning green stage to promote the growth. In order to avoid the damage on the growth of crops because of late spring coldness, people will do jobs of coldness protection, freeze protection, pest management and winter wheat weeding. In Sichuan region, local people will seize time digging the seedbed of early rice, and complete the tasks of seed selection, seed drying in sunshine and field management for summer harvesting crops. As the saying goes, when the rain of the Beginning of Spring arrives, people must get up early and go to bed late. In the Beginning of Spring, the spring ploughing and spring sowing should be in full swing, and the weather will be given close attention for the planting of early rice.

为了迎接春天的到来，以提醒人们及时进行春耕春种和食材生产，自先秦时起，立春日的迎春就成了一项重要的民俗活动，也是历代帝王和普通百姓都要参加的迎春庆贺礼仪。从一定意义上讲，立春日迎春的民俗就是促进食材生产的仪式和活动。在四川地区，清朝和民国地方志中都有关于“迎春”“点春”“春台”“打春”等仪式流程的记载，从中颇能看出四川人对立春这一节气的重视。据清代嘉庆二十一年（1861年）的《华阳县志》记载：“‘立春’前一日，府尹率县令、僚属迎春于东郊，仪仗甚盛，鼓乐喧阗，芒神、土牛导其前，并演春台，又名高妆社火。士女骈集，谓之‘看春’。次日，鞭土牛于府署，谓之‘打春’。”资阳地区与华阳县同属成都平原，在立春日历来也有迎春、打春等民俗活动。

In order to welcome the spring and remind people to plant crops and produce food ingredients timely, spring heralding ceremony has become an important folk custom activity since pre-Qin times (before 221 BC). Both emperors and common people must take part in this ceremony. To some extent, this folk custom activity is a ceremony and activity which supposes to promote the producing of food ingredients. In Sichuan region, you can find the records of the related ceremonies' process, like Ying Chun, Dian Chun, Chun Tai, Da Chun, etc. in local chronicles from Qing Dynasty and the Republic of China era. Those records have shown the importance of the Beginning of Spring in Sichuan people's mind. According to the *Huayang County's Chornicles* in the 21st year of Jiaqing Emperor (1861), the chief of local government led all his staffs to welcome the spring in eastern suburbs one day before the Beginning of Spring. The ceremony was great with loud music filling the air. People put the clay ox and the statue of spring god (Mang Shen) ahead, and performed local opera in the fields. Ladies had gathered to watch ceremony (Kan Chun). On the day of the Beginning of Spring, the clay ox was beaten in the government department, which are called Da Chun. Both Ziyang region and Huayang County are located in Chengdu Plain, and have the same folk custom activities on the day of the Beginning of Spring.

饮食养生 | DIETARY REGIMEN

立春时节，四川地区阳热渐生而阴寒未尽，自然界正处于阴退阳进、寒去阳回的交替时期，天气乍暖还寒，气温忽高忽低，人体易受风寒之扰。尤其在初春，此时人体阳气始发，新陈代谢增强，如能在春季大自然“发陈”之时，借阳气上升、人体新陈代谢旺盛之机科学养生，对健身防病都十分有利。具体而言，立春前后适合多食具有辛甘发散性质的食物，四川民间尤其提倡常食“芽菜”。所谓“芽菜”，是泛指食材的嫩芽，如春笋、姜芽、香椿芽、黄豆芽、绿豆芽等，食芽菜可使人体阳气得以生发。杜甫诗言：“夜雨剪春韭，新炊间黄粱。”春初早韭不仅鲜嫩可口，而且辛温散寒，对于宣发阳气有一定益处。此外，春季适宜养肝，可适当食用利于护肝的食材，使肝气得以疏泄条畅，达到气血充沛、身体健康的养生目的。如选择动物肝脏、绿色蔬菜等食材与菊花、枸杞子等搭配组成食疗方，能平肝阳、滋肝阴，与红枣、桂圆等配伍，还可补养肝血。

In the Beginning of Spring, Sichuan region is still cold although the temperature starts to rise slowly. Unstable weather and temperature will cause colds easily, especially in the early spring, the body's metabolism will speed up. We can take good use of the natural moment to cure diseases and preserve health scientifically. To be more specific, it is better to eat more foods in symplectic nature around the Beginning of Spring. In Sichuan folk, people are encouraged to eat sprout. Sprout here refers to the new shoots of plants, like bamboo shoot in spring, ginger shoot, Chinese toon sprout, soybean sprout, mung bean sprout, etc. Eating sprout can strengthen the yang qi (positive life-energy) in human body. There is famous poem by Du Fu,"The night rain like the scissor cutting Spring leeks, the cooking smoke is raising around beams of the house." Spring leeks in the early spring is not only delicious, but also with the function of dispel cold, which will help to produce the yang qi in this season. Moreover, it is suitable to nourish the liver in spring. We should eat some food which can protect the liver and dredge the liver to keep us in good health. For example, we can choose dietary therapy prescription composed of animal liver, green vegetable, chrysanthemum and fructus lycii, which are able to nourish the liver. Adding red date and longan can enrich the liver blood as well.

立春美食 | FINE FOODS IN THE BEGINNING OF SPRING

立春的主要食俗是“咬春”，人们普遍认为，“咬春”可以除却春困。清朝潘荣陛在《帝京岁时记》中对“咬春”有如下之言：“新春日献辛盘。虽士庶之家，亦必割鸡豚，炊面饼，而杂以生菜、青韭芽、羊角葱，冲和合菜皮，兼生食水红萝匐，名曰咬春。”文中的“辛盘”，又称之为“五辛盘”，据《本草纲目・菜部》言：“五辛菜，乃元旦、立春以葱、蒜、韭、蓼蒿、芥辛嫩之菜杂和食之，取迎新之意，谓之五辛盘。”服食五辛，可杀菌驱寒。此外，立春之日，用蔬菜、水果、饼饵等装盘为食或馈赠亲友，谓之“春盘”，古时的帝王也会在立春前一天，以春盘并酒赏赐近臣，这是立春之时流行于我国古代的另一种习俗，兴起于晋朝，到唐宋时，吃春盘之风日渐盛行，《唐四时宝镜》言：“立春日，食芦菔、春饼、生菜，号春盘。”唐朝沈佺期《岁夜安乐公主满月侍宴》诗云：“岁炬常然桂，春盘预折梅。”另有金代元好问《春日》诗云：“里社春盘巧欲争，裁红晕碧助春情。”

The main eat custom in the Beginning of Spring is biting the spring (Yao Chun). People believe that biting the spring can deal with spring drowsiness. Pan Rongbi in Qing Dynasty has described the biting spring in his book *Record of Customs in the Capital of Empire* as follows, "Normal people will prepare pungent food during

the Beginning of Spring. All kinds of families must cook chicken and pork, make flatbread, and eat with lettuce, leek shoot, spring onion, radish. People call it biting the spring." Pungent food is also called Five Xin Pan in China. *Compendium of Materia Medica* says that Five Xin dishes mean eating green onion, garlic, leek, artemisia and mustard leaf together on New Year's Day and in the Beginning of Spring. People call it Five Xin Pan to welcome the arrvial of the new year. Eating Five Xin dishes can kill germs and take the chill off. Furthermore, on the day of the Beginning of Spring, people will put vegetables, fruits and cookies on plate for eating or sharing with friends, which is called Spring Plate. Ancient emperors will give Spring Plate and liquor to courtiers one day before the Beginning of Spring. This custom has begun in Jin Dynasty (266-420) and become populous in Tang Dynasty (618-907) and Song Dynasty (960-1279). The book *Four Seasons' Collection of Tang Dynasty* said, "People eat radish, spring pancake and lettuce on the day of the Beginning of Spring. It is called Spring Plate." Shen Quanqi in Tang Dynasty and Yuan Haowen in Jin Dynasty (1115-1234) have mentioned about Spring Plate separately in their poems. There is poem from *Princess Anle's Full-Moon-Birth Banquet on the New Year's Eve*, "Time flies just like a candle burns, the guests are enjoying the fruits and cookies on plate and will go to see the plum blossom." The poem *Spring Day* by Yuan Haowen in Jin Dynasty tells, "People are preparing plump pungent food, ladies are making vivid tailor which liven the Spring in the air."

春卷，也是古代立春的传统美食。据宋朝陈元靓《岁时广记》记载，京师富贵人家造面茧，以肉或素做馅。又因为是在立春之日为此，所以又称为“探春茧”。其形状像蚕茧一样的面茧，即今天的“春卷”。在宋朝，春卷有米薄皮春茧、子母春茧、活糖沙馅诸色春茧等多个品种，到明清时期已成为全国各地著名的立春美食。在资阳乃至整个四川，春卷都是立春日，也是春天来临的标志性美食。

Spring roll is a traditional food for the Beginning of Spring. According to Chen Yuanliang's *Sui Shi Guang Ji* in Song Dynasty, rich families in capital made flour cocoon with vege and non-vege fillings which called Mian Jian in China. Because people made it on the day of the Beginning of Spring, it has been called Tan Chun Jian as well. It's shaped like a cocoon and is the origin of today's spring roll. In Song Dynasty, there were rice surface spring roll, double spring roll, sweet filling spring roll, etc. In Ming Dynasty and Qing Dynasty, spring roll has become a nation-wide famous food. In Ziyang even the whole Sichuan province, spring roll is the iconic food for the day of the Beginning of Spring and the coming of spring.

如意春盘

Ruyi Spring Plate

食材配方

卤牛肉50克　基围虾50克　三文鱼50克　芦笋20克　小黄瓜20克　樱桃萝卜15克　西兰花15克　胡萝卜15克　白萝卜20克　青笋10克　巧克力酱5克　食盐3克　泡菜汁水100毫升　葱油10毫升

制作工艺

① 基围虾用沸水煮熟后去壳；白萝卜切成约0.1厘米厚的片，用泡菜汁水浸泡15分钟；西兰花、芦笋分别放入沸水锅中煮熟后晾凉；胡萝卜切成细丝。

② 胡萝卜丝用白萝卜卷成珊瑚萝卜；三文鱼先切成约0.3厘米厚的片，再制成花形；小黄瓜、樱桃萝卜、卤牛肉分别切成约0.1厘米厚的片。

③ 将切好的原料依次按照小黄瓜、樱桃萝卜、卤牛肉、珊瑚萝卜、基围虾、三文鱼、芦笋、西兰花摆成田野图案，再用青笋制成的柳枝做点缀，最后用巧克力酱题字，刷上葱油即可。

风味特色

色泽鲜艳，刀工细腻，味型多样，造型美观大方。

创意设计

春盘是我国古代立春之日的节气美食，也是春天来临的标志性美食。此菜将多种原料用多种烹饪方法，调以不同味型并组成图案，充分体现出刀工、造型与味型之美，寓意春意盎然、万物欣欣向荣。在养生方面，牛肉补脾益气，卤制后生阳养阳效果较好，基围虾补肾壮阳，三文鱼益气开胃，三者搭配尤宜阳气生发且顾护脾胃。同时，搭配蔬菜同食，可补充膳食纤维、维生素及矿物质，营养更丰富。

Ingredients

50g flavored beef (boiled in Sichuan-style broth); 50g greasyback shrimps; 50g salmon; 20g asparagus; 20g small cucumbers; 15g cherry belle radishes; 15g broccoli; 15g carrots; 20g white radishes; 10g Chinese lettuce stems; 5g chocolate sauce; 3g salt; 100ml pickle brine; 10ml spring onion oil

Preparation

1. Boil the greasyback shrimps till fully cooked and shell. Cut the white radishes into 0.1cm thick slices, and soak in pickle brine for 15 minutes. Boil the broccoli and asparagus till fully cooked and leave to cool. Shred the carrots.
2. Roll up the carrot slivers with white radish slices. Slice the salmon into 0.3cm thick slices, and arrange the slices like a flower. Cut the small cucumbers, cherry belle radishes, and flavored beef into 0.1cm thick slices separately.
3. Lay on a plate small cucumbers, cherry belle radishes, flavored beef, white radish rolls, shrimps, salmon, asparagus and broccoli to present a spring field view, then garnish with the lettuce stem slivers as willows. Write with chocolate sauce, and brush with spring onion oil.

Features

bright color; fine cutting; multiple flavors; beautiful presentation

Notes

Spring Plate, a traditional festive food for the Beginning of Spring since ancient times, marks the commencement of spring. With various ingredients, varied culinary methods, multi-flavors and beautiful presentation, this dish symbolizes that the land is bursting with life. In food therapy, beef has the nature of replenishing qi and increasing the appetite. It can better nourish yang after being boiled and flavored in broth. Greasyback shrimps can tonify the kidney and strengthen yang. Salmon can replenish qi and arouse the appetite. Combining these three promotes qi circulation and benefits the stomach and spleen. Meanwhile, vegetables provide more balanced nutrition, for they contain dietary fibers, vitamins and minerals.

韭黄烘蛋

Fried Omelet with Yellow Chives

食材配方

鸡蛋液200毫升　韭黄50克　胡萝卜50克　葱花30克　食盐4克　水淀粉20克
面粉30克　食用油30毫升

制作工艺

① 韭黄、胡萝卜分别切成碎粒。
② 鸡蛋液入碗，放入食盐、水淀粉、面粉、韭黄、胡萝卜、葱花搅拌均匀。
③ 平底不粘锅置火上，入油烧至120℃时，倒入鸡蛋液煎至底面酥香、表皮凝固后，再翻面煎至表皮金黄、酥香，出锅后切成方块，装入盘中即成。

风味特色

色泽金黄，质地鲜嫩酥软，味道咸鲜香浓。

创意设计

立春前后宜多食辛甘发散性质的食物，四川民间尤其常食“芽菜”，韭黄便是其中极具代表的特色食材之一。此菜将韭黄与鸡蛋、胡萝卜搭配成菜，营养均衡、色彩鲜艳、香味浓郁。在养生方面，韭黄温中行气，能抗氧化，搭配养肝明目的胡萝卜与滋阴养胃的鸡蛋，温阳兼能养阴，非常适合此时节的养生需要。

Ingredients

200ml beaten eggs; 50g yellow chives; 50g carrots; 30g spring onions, finely chopped; 4g salt; 20g average batter; 30g wheat flour; 30ml cooking oil

Preparation

1. Chop the yellow chives and carrots into grains separately.
2. Combine the beaten eggs, salt, batter, wheat flour, yellow chives, carrots and spring onions in a bowl, and blend well.
3. Heat oil in a frying pan to 120℃, add beaten eggs and fry till the bottom side is golden brown and aromatic and the surface is set. Flip the omelet upside down and fry till the other side is golden brown, then remove. Cut into rectangular pieces, and transfer to a serving dish.

Features

golden brown in color; tender texture; savory taste; lingering aroma

Notes

During the Beginning of Spring period, it is customary to eat pungent foods. Local people in Sichuan prefer having yacai (sprouts), and yellow chives are one of the typical kinds. This dish uses as the main ingredients yellow chives, eggs and carrots, which are not only nutritious but have bright colors and pleasant aromas as well. As for their functions in food therapy, yellow chives contain antioxidant, warm the stomach and spleen, and smooth qi circulation. Carrots contain substances to improve the vision and nourish the liver. Eggs can replenish yin and nourish the stomach. Combining three ingredients together, this dish can warm yang and nourish yin at the same time, which is just the right food for this season.

资阳春卷

Ziyang Spring Rolls

食材配方

春卷皮500克　胡萝卜200克　青笋200克　黄瓜200克　绿豆芽200克　紫甘蓝200克　食盐15克　复制酱油50毫升　食醋13毫升　味精1克　芝麻油5毫升　红油辣椒75毫升　葱花100克

制作工艺

① 青笋、紫甘蓝、黄瓜和胡萝卜分别切成细丝，放入凉开水中浸泡待用；绿豆芽入沸水锅中焯熟后捞出，放入清水中漂冷备用。

② 在春卷皮上分别放入胡萝卜丝、青笋丝、紫甘蓝丝、黄瓜丝和绿豆芽，先将其卷成圆筒，再切成两段放入盘中。

③ 将食盐、复制酱油、红油辣椒、味精、食醋、芝麻油、葱花放入小碗中调匀成味汁，淋在春卷上即成。

风味特色

皮薄绵软，馅心鲜艳，味道酸辣回甜，清香可口。

创意设计

春卷是中国民间立春时节的传统美食。此菜在传统淋味春卷制法的基础上，以资阳出产的青笋、胡萝卜、黄瓜、绿豆芽及紫甘蓝为馅，既让春卷馅心色泽艳丽，也增加了其营养和食疗价值。其中，绿豆芽生发阳气，配合青笋、黄瓜等养肝护肝，春卷皮、胡萝卜健脾益气，调以红油辣椒则辛散温通。此外，蔬菜富含维生素、矿物质及膳食纤维，紫甘蓝富含花青素，有利于抗氧化、降低“三高”。

Ingredients

500g spring roll wrappers; 200g carrots; 200g Chinese lettuce stems; 200g cucumbers; 200g mung bean sprouts; 200g red cabbages; 15g salt; 50ml compound soy sauce; 13ml vinegar; 1g MSG; 5ml sesame oil; 75ml chili oil; 100g spring onions, finely chopped

Preparation

1. Shred the lettuce stems, red cabbages, cucumbers and carrots respectively then soak in cold boiled water. Blanch mung bean sprouts, remove and soak in cold boiled water to cool.

2. Place slivers of carrots, lettuce stems, red cabbages, cucumbers and mung bean sprouts on spring roll wrappers, roll each wrapper into a cylinder, halve and transfer to a serving dish.
3. Mix the salt, compound soy sauce, chili oil, MSG, vinegar, sesame oil, spring onions in a small bowl to make the seasoning sauce. Drizzle the sauce over the spring rolls.

Features

thin wrappers; colorful stuffing; spicy and sour tastes with a lingering sweet flavor; fresh fragrance

Notes

Spring rolls are a traditional Chinese food during the Beginning of Spring. Using a variety of vegetable as the main ingredients, this dish is not only appealing to the eye but rich in nutrients as well. In terms of food therapy, mung bean sprouts can produce yang. Lettuce and cucumber are good for the liver. Spring roll wrappers and carrots can fortify the spleen and tonify qi. Red chili oil is pungent and helps remove inner cold. Additional, vegetables are rich in vitamins, minerals and dietary fibers. Red cabbage is famous for its anthocyanidin, an antioxidant that can lower blood sugar, blood pressure and cholesterol.

RAIN WATER

雨水

春风化雨　甘露时降

春夜喜雨

唐·杜甫

好雨知时节，当春乃发生。
随风潜入夜，润物细无声。
野径云俱黑，江船火独明。
晓看红湿处，花重锦官城。

雨水物候 | PHENOLOGY IN RAIN WATER

雨水是二十四节气中的第二个节气。当太阳到达黄经330° 时为雨水，时间在每年的2月18日～20日。元朝吴澄《月令七十二候集解》言："雨水，正月中。天一生水，春始属木，然生木者，必水也，故立春后继之雨水。且东风既解冻，则散而为雨水矣。"意思是春天草木生长需要雨水滋润，而春风令冰雪消融，带来雨水。雨水节气即指降雨开始、雨量渐增。

Rain Water is the second of the 24 Solar Terms. It begins when the sun reaches the celestial longitude of 330°. It usually begins from 18th to 20th February every year. In his book *A Collective Interpretation of the Seventy-Two Phenological Terms*, Wu Cheng (Yuan Dynasty 1271-1368) said that the growth of plants in spring need rain, and the spring wind will bring rain by melting ice and snow. Rain Water means the start of raining in this season, and the increasing of rain.

中国古代把雨水后的十五天分为三候:一候獭祭鱼；二候鸿雁北；三候草木萌动。雨水后第一个五天，水獭开始捕鱼了，将鱼摆在岸边如同先祭后食的样子；第二个五天，南雁北飞；第三个五天，在春雨润物无声中，草木开始抽芽发叶。俗语有言，"春雨贵如油"。雨水相应的花信为：一候菜花；二候杏花；三候李花。

Ancient China has divided the 15 days after Rain Water into 3 pentads: in the first pentad, otters start to catch fish and put the fishes along the river bank as ritual first; in the second pentad, wild geese fly from the south to the north; in the third pentad, plants begin to sprout and grow leaves. As proverb goes, "rain in spring is as rich as oil". The news of flowers blooming in Rain Water are as follows: rape flower for the first pentad, almond flower for the second pentad and plum blossom for the third pentad.

食材生产 | PRODUCING FOOD INGREDIENTS

雨水时至，雨量渐渐增多，有利于越冬作物返青、生长，人们常常在此刻抓紧时间进行越冬作物的田间管理和春耕春播工作。民间普遍认为，雨水这一天的雨是丰收的预兆，许多地区都流行一句农谚："雨水有雨庄稼好，大春小春一片宝。"资阳地区更有农谚道，"立春雨水，空土挖来焊起"，以此使土地进行短暂的休养生息，可为食材生长奠定较好的种植基础。

In Rain Water, increasing of rain fall gradually is helpful for the overwintering crops to turn green and resume growth. People usually seize time managing the filed and starting spring ploughing and spring sowing. According to the folk belief, the rain on the day of Rain Water is the omen of good harvest. There is a farmer's proverb in lots of regions saying, "rain in Rain Water will help the crops, and it means a bright future almost through the year". A farmer's proverb in Ziyang region says that the rain in Rain Water make the soil rest. The short rest can provide a good basis to the growth of food ingredients.

饮食养生 | DIETARY REGIMEN

雨水时节，四川地区的天气阴晴不定，时冷时热、乍暖还寒的气温波动，不仅会左右已经萌芽和返青的食材生长，也对人体健康有较大影响。由于自然界的春季与人体五脏中的肝脏相对应，均归属“木”行，因此，人们在春季容易肝气过旺，从而对隶属“土”的脾胃功能产生不良影响，妨碍食物的正常消化和吸收，且酸性收敛，不利于肝气的畅达。唐朝医学家、养生家孙思邈提出：春日宜省酸增甘，春时宜食粥。在五味之中，因为酸味入肝脏能补肝，甘味入脾脏能补脾，雨水时节容易导致肝气过旺而克制脾胃的功能，所以，此时要少吃酸味食材，以免肝脏功能偏亢，降低其疏泄功能，而应适当吃一些甘味食材，如山药、红枣、小米、糯米、薏苡仁、黄豆、胡萝卜、红薯、土豆、桂圆、栗子等，以增强人体的脾胃功能。同时，宜少食生冷油腻、难以消化的食物，可适当多食用谷米熬制，且利于消化、吸收的粥，以养护和补益脾胃。若结合雨水节气寒湿偏多的特点，粥中可选加薏苡仁、白扁豆、芡实，或高良姜、桂圆肉等健脾、散寒除湿之物，效果更好。

In Rain Water, the weather and temperature in Sichuan region are not stable. Even when it is warmer there is still a chill. Such climate is harmful for the growth of plants and human health. According to the TCM theory, spring is in correspondence to the human liver, and both belong to the wood primary element. Hence in Spring, the qi in liver is very strong to affect the spleen which belongs to the earth primary element. Noraml digestion and absorption of food will meet difficulties, and the qi in liver will be blocked too. Sun Simiao, great medical scientist from Tang Dynasty, pointed out that people should eat less sour food and more sweet food, and it was a proper time to eat porridge in spring. Among the five flavors, sour flavor can tonify the liver and sweet flavor can tonify the spleen. In Rain Water, the liver is so strong to hinder spleen, therefore we should eat less sour food and more sweet food, such as Chinese yam, red date, millet, glutinous rice, coixenolide, soy bean, carrot, sweet potato, potato, longan, chestnut and so on, which can improve the function of spleen and stomach. In order to protect and tonify spleen and stomach, people should eat less cold, oily and hard-digested food, and eat more porridge made of cereals properly. If we take the cold-dampness of Rain Water into consideration, we should add coixenolide, white hyacinth bean, gorgon fruit or galangal, longan pulp into the porridge, which can dispel cold and remove dampness better.

雨水美食 | FINE FOODS IN RAIN WATER

在四川一些地区，雨水节气有与人生礼俗相关的活动和美食。在雨水这一天，孩子的父母为孩子“拉保保”，取“雨露滋润易生长”之意。“保保”是四川方言，即干爹的意思。找干爹的目的则是借助干爹的福气来荫庇孩子，让儿女顺利、健康地成长。在雨水当天，准备拉保保的父母带着孩子，提着酒菜和香蜡钱纸到特定场所，在人群中找准孩子的干爹对象后连声说“打个干亲家”，便摆好酒菜、焚香点蜡，让孩子“拜干爹，叩头”，同时，请干爹喝酒吃菜、给孩子取名字。当这些程序完成后，拉保保就算成功了。另外，雨水这一天，川西地区生育了孩子的妇女还要带上“罐罐肉”等礼物回娘家。将用砂锅炖好的猪蹄、黄豆、海带等装罐，再用红纸、红绳密封罐口，即为罐罐

肉。将其带回娘家，是借此表达将自己辛苦养育成人的父母的感谢和敬意。

In some parts of Sichuan, Rain Water has special activities and fine foods related with etiquette and custom. On the day of Rain Water, parents will seek godfathers for their children, which is called La Baobao. People choose that day to borrow the magic of rain which can nourish the growth. Baobao comes from Sichuan dialect and means godfather. The purpose of seeking godfather is to protect and bring welfare to the children, then children can grow up healthily. On the day of Rain Water, parents who plan to seek the godfather will bring their children, dishes, liquor and ritual objects to the specific site. There, they will say "lets be relatives" to the chosen man, and ask the children to kowtow in front of the ritual objects. The newly chosen godfather will be invited to enjoy liquor and dishes, and give the children new names. When all the process finish, La Baobao is done. Moreover, on the day of Rain Water, women who have given birth to children in west Sichuan will bring gifts like Potted Pork to her parents' home. As for the Potted Pork, pig trotters are cooked with soybean and kelp in casserole which has been sealed with red paper and thread. Bringing this gift back home is to show her sincere appreciation and respect to parents who have raise her up.

雨水节气的时间，在许多年份又与农历正月十五元宵节临近。四川民间过元宵节有吃汤圆的习俗。清朝吴德纯《锦城新年竹枝词》言："食品元宵巧制难，浮圆甘美簇春盘。佳名爱取团圞意，笑指郎君仔细看。"元宵，在四川又名汤圆，取团圆之意。

In history, the date of Rain Water was close to the Spring Lantern Festival in many cases. Sichuan people usually eat glue pudding (Tangyuan) in Spring Lantern Festival. Wu Dechun in Qing Dynasty (1636-1912) has mentioned the glue pudding in his poem, "glue pudding is difficult to make，each one is sweet and beautiful in plate. It's name means reunion，gentleman should understand this connotation." In Sichuan, glue pudding is called Tangyuan too, means reunion.

罐罐肉 | Potted Pork

食材配方

带皮五花肉500克　笋子250克　姜片15克　葱段20克　糖色25克　东坡红酱油5毫升　食盐8克　料酒30毫升　八角2克　干辣椒2克　干花椒1克　香叶1克　清水1000毫升　食用油30毫升

制作工艺

① 五花肉切成约2厘米见方的块；笋子切成滚刀块。

② 锅置火上，入油烧至150℃时，放入五花肉、姜片、葱段煸炒出香味，入笋子、料酒、糖色、东坡红酱油、食盐、八角、干辣椒、干花椒、香叶、清水，烧沸后改用小火煨2小时，然后去掉姜、葱及辛香料，装入瓷器罐罐中，继续用小火煨至肉质软烂，汤汁浓稠时即成。

风味特色

成菜色泽红亮，肉质软糯，味道咸鲜醇厚。

创意设计

雨水时节，在四川民间家庭中，晚辈常常烹制“罐罐肉”以孝敬长辈，传承和体现着“敬老孝亲”的优秀文化传统。此菜在传统罐罐肉制法的基础上加以改进，成菜软糯、香醇，营养丰富，老少皆宜。此外，猪肉滋阴养肾，春笋化痰消胀，二者互补，且春笋中含有大量膳食纤维，有助于降低“三高”。

Ingredients

500g pork belly with skin attached; 250g bamboo shoots; 15g ginger, sliced; 20g spring onions, segmented; 25g caramel; 5ml Dongpo dark soy sauce; 8g salt; 30ml Shaoxing cooking wine; 2g star aniseeds; 2g dried chilies; 1g Sichuan peppercorns; 1g bay leaves; 1,000ml water; 30ml cooking oil

Preparation

1. Chop the pork belly into 2cm cubes. Cut the bamboo shoots diagonally.
2. Heat oil in a wok to 150℃, add the pork, ginger, spring onions and stir fry till aromatic. Add bamboo shoots, cooking wine, caramel, dark soy sauce, salt, star aniseeds, dried chilies, Sichuan peppercorns, bay leaves and water, bring to a boil, turn down the heat and simmer for 2 hours. Remove the ginger, spring onions and spices, then transfer to a porcelain pot. Simmer till the pork is soft and the sauce is thick.

Features

reddish brown in color; soft and glutinous pork; savory taste and lasting fragrance

Notes

During Rain Water period, Sichuan people usually show their filial piety to family seniors by cooking Potted Pork for them. This custom embodies the Chinese tradition of respecting the elders. This soft, glutinous and savory dish is suitable for both young and old. Additionally, pork can nourish yin and replenish the kidney, while spring bamboo shoots can resolve phlegm and promote digestion. They complement each other in food therapy. Besides, spring bamboo shoots have a lot of dietary fibers, which are helpful to reduce blood sugar, blood pressure and cholesterol.

养生三元烩

Three Health Treasures

食材配方

铁棍山药150克　红腰豆60克　薏仁50克　浓汤200毫升
金瓜汁50毫升　食盐3克　水淀粉20克

制作工艺

① 铁棍山药制净后切成长约3厘米的段。
② 将铁棍山药、红腰豆、薏仁分别放入蒸箱内蒸制成熟后取出，摆放在盘中。
③ 锅置火上，入浓汤、食盐、金瓜汁烧沸，最后用水淀粉收汁，起锅淋在食材上即成。

风味特色

成菜色泽金黄，味道咸鲜，营养丰富，造型独特。

创意设计

雨水时节，雨量增多，气候开始变得潮湿，可适当食用有利于散寒祛湿、增强脾胃功能的食物。此菜将薏仁、山药和红腰豆三种食物合烹而成，散寒除湿效果良好。其中，山药健脾养胃、生津益肺，此时食用，既暗合“省酸增甘”之意，也具有调节免疫功能及抗氧化、延缓衰老的作用，再搭配以补肾益气的红腰豆和利水健脾的薏仁，更能祛湿益气。

Ingredients

150g iron stick yams; 60g red kidney beans; 50g coix seeds; 200ml thick stock; 50ml pumpkin juice; 3g salt; 20g average batter

Preparation

1. Clean the yams and cut into 3cm long sections.
2. Steam the yams, red kidney beans and coix seeds in a steamer till cooked through, remove and transfer to a serving dish.
3. Heat a wok, add the stock, salt, pumpkin juice and bring to a boil. Add the batter to thicken the sauce, then drizzle over the yams, beans and coix seeds.

Features

golden in color; savory taste; rich nutrition

Notes

The weather becomes humid during Rain Water period, when it is suitable to have foods that can dispel dampness and fortify the spleen and stomach. Combining coix seeds, yams and red kidney beans together, this dish has better functions of dispelling dampness and coldness of the body. Yam is especially good for nourishing the spleen, increasing appetite, inducing saliva production and benefiting the lung. It also helps with immunity and has anti-aging and anti-oxidation effects. Red kidney beans nourish the spleen and replenish qi. Coix seeds promote body water excretion and strengthen the spleen.

山药汤圆 | Yam Stuffed Tangyuan

食材配方

糯米粉270克　清水200毫升　铁棍山药150克　糖粉15克　熟面粉10克　猪油20克　蜜饯丝10克

制作工艺

① 将糯米粉用清水调制成糯米面团。

② 铁棍山药去皮，切段后蒸熟，趁热压成泥，加入糖粉、熟面粉调匀，待晾凉后加入猪油、蜜饯丝搅拌均匀，制成山药馅。

③ 取13克糯米面团，包入7克山药馅，制作成汤圆生坯。

④ 锅置火上，入水烧沸，放入汤圆煮至成熟后捞出装入碗中即成。

风味特色

色泽洁白，皮薄馅多，细腻爽滑，味道香甜。

创意设计

山药是雨水时节的最佳养生食材之一。本款汤圆将四川民间常用的芝麻馅换作山药馅，更适合此节气食用。其中，糯米温中散寒，山药健脾养胃，二者结合，可共收健脾调中之效，是春季“省酸增甘”的食用佳品。此外，山药含有薯蓣皂苷、糖蛋白、多糖等成分，有降低血糖、提高免疫力、抗氧化、抗衰老等作用。

Ingredients

270g glutinous rice flour; 200ml water; 150g Chinese yams; 15g caster sugar; 10g pre-cooked wheat flour; 20g lard; 10g candied fruits, shredded

Preparation

1. Mix the glutinous rice flour with water to make glutinous rice flour dough.
2. Peel the Chinese yams and cut into sections. Steam the sections in a steamer till cooked through, and mash while they are still hot. Blend in the caster sugar and pre-cooked wheat flour and mix well. Leave to cool. Add lard and candied fruits to make the stuffing.

3. Wrap up 7g stuffing with 13g glutinous rice flour dough to make Tangyuan.
4. Heat water in a wok, bring to a boil and roll in the Tangyuan. Boil till cooked through, and transfer to a serving bowl.

Features

snow white color; thin wrapper and rich fillings; soft and smooth texture; sweet and succulent tastes

Notes

Chinese yam is one of the best health foods during the time of Rain Water. Therefore, this dish uses yam as stuffing instead of sesames which are a more common filling for Tangyuan (stuffed rice balls) in Sichuan. In terms food therapy, glutinous rice is mild in nature and helps to expel inner cold while Chinese yam helps to tonify the spleen and nourish the stomach. Combining the two ingredients, this dish has functions of nourishing the spleen and regulating the stomach. Furthermore, it is an ideal dish for this season, for it is accorded with the traditional dietary principle of eating more sweet but less sour food in spring. In addition, the Chinese yam contains dioscin, glycoprotein and polysaccharide, which are beneficial for immunity enhancement, anti-oxidation and anti-aging.

THE WAKING OF INSECTS

惊蛰

阳和启蛰品物皆春

闻雷

唐·白居易

瘴地风霜早，
温天气候催。
穷冬不见雪，
正月已闻雷。
震蛰虫蛇出，
惊枯草木开。
空馀客方寸，
依旧似寒灰。

惊蛰物候 | PHENOLOGY IN THE WAKING OF INSECTS

惊蛰是二十四节气中的第三个节气。惊蛰古称“启蛰”，标志着仲春时节的开始，此时太阳到达黄经345°，时间在每年的3月5日～7日。元朝吴澄《月令七十二候集解》言：“惊蛰，二月节”，“是蛰虫惊而出走矣”。在这一节气之前，动物入冬藏伏土中，不饮不食，称为“蛰”。到“惊蛰”之时，古人认为春雷惊醒了蛰居的动物。此时，天气转暖，渐有春雷，蛰虫惊醒出走。

The Waking of Insects is the third of the 24 Solar Terms. The Waking of Insects, which was called Qi Zhe in ancient times, begins when the sun reaches the celestial longitude of 345°. It usually begins from 5th to 7th March every year and means the start of the mid-spring. In his book *A Collective Interpretation of the Seventy-Two Phenological Terms*, Wu Cheng said that animals hibernate underground before this term, which is called Zhe. The spring thunder awakens those animals in the Waking of Insects. In this period, it is getting warm and the insects begin to move because of spring thunder.

中国古代将惊蛰后的十五天分为三候:一候桃始华；二候仓庚鸣；三候鹰化为鸠。蛰伏了一冬的桃花开始开花，并逐渐繁盛；仓庚即黄鹂鸟，它们感知到春天的气息，发出婉转悦耳的啼鸣；动物开始繁殖，鹰开始躲起来繁育后代，而原本蛰伏的鸠开始鸣叫求偶，古人没有看到鹰，却看到周围的鸠好像一下子多起来，就误以为是鹰变成了鸠。惊蛰相应的花信为：一候桃花；二候棣棠；三候蔷薇。

Ancient China has divided the 15 days after the Waking of Insects into 3 pentads: in the first pentad, peach blossoms; in the second pentad, orioles who can feel the smell of spring begin to sing; in the third pentad, eagles has become turtle doves. In the Waking of Insects, eagles hide to breed the descendants and turtle doves sing for courtship. The ancient people could not see eagles, but more turtle doves, therefore they believed that eagles become turtle doves. The news of flowers blooming in the Waking of Insects are as follows: peach flower for the first pentad, kerria japonica for the second pentad and rosa multiflora for the third pentad.

食材生产 | PRODUCING FOOD INGREDIENTS

“到了惊蛰节，锄头不停歇。”我国人民自古便将惊蛰视为春耕的重要日子，大部地区进入春耕大忙季节。此时，四川地区春暖融融，最适合春季作物和蔬菜的播种，如春玉米、春大豆、南瓜、菜瓜、早毛豆、菜豆、豇豆等。随着气温回升，茶树也开始渐渐萌芽，人们及时修剪并追施“催芽肥”，以促进其多分枝、发叶，提高茶叶产量。为了促进食材的顺利生长，减少害虫危害，民间出现了一些特殊的风俗，如与四川接壤的陕西一些地区，人们要吃炒豆，是将黄豆用盐水浸泡入锅爆炒后食用，此时的黄豆象征害虫。

“In the Waking of Insects, you can not stop hoeing”. Chinese people consider the Waking of Insects an important day for spring ploughing. Lots of regions are in busy season. Sichuan region is warm in this period and suitable for sowing spring crops and vegetables, like spring corn, spring soybean, pumpkin, snake melon, early edamame, french bean, cowpea, etc. As temperature rises, tea trees begin to sprout. People will prune the branches and feed the trees to increase the production of tea. In order to promote the food ingredients to grow healthily and reduce the damage of pest, some special customs have appeared. For example, people eat stir-fried soybean in Shaanxi next to Sichuan. Soybean stands for pest. Soak the soybean in brine first and then fry it.

饮食养生 | DIETARY REGIMEN

惊蛰后天气转暖，四川地区气温回升较快，但冷空气活动仍较频繁，有时会出现“倒春寒”现象，人们需要根据天气冷暖变化及时增减衣服，预防感冒、咳嗽等外感疾病的发生。进入惊蛰以后，随着天气转暖，人们也常会感到困倦无力、昏昏欲睡，民间称之为“春困”。为预防外感疾病和春困，要适当增加营养，可适度吃一些含有“辛”味的应季食物，如韭菜、洋葱、香椿等辛香蔬菜。惊蛰时节仍然要贯彻“省酸增甘”的饮食养生原则，饮食宜清淡、稍温，贵精不贵多，多食用新鲜蔬菜及蛋白质丰富的食物，如春笋、菠菜、芹菜、鸡肉、鸡蛋、牛奶等，少吃辣椒、花椒、胡椒等过于辛辣刺激及生冷油腻的食物，既能保证肝气生发、肝血充沛，也可避免过食肥甘厚味、助阳上火而损伤脾胃。

It is getting warm after the Waking of Insects. The temperature of Sichuan rises quickly, but sometimes late spring coldness occurs. People need change clothes according to the weather to prevent influenza and cough. After the Waking of Insects, people usually feel sleepy and weak, which is called spring drowsiness. In order to avoid illness and spring drowsiness, we should get more nutrition and eat some seasonal pungent food, like leek, onion, toona sinensis, etc. In the Waking of Insects, people still need follow the principle of eating less sour and more sweet. We should eat light food. Fresh vegetables and food full of protein, like bamboo shoots in spring, spinach, celery, chicken, egg, milk are welcome. Spciy and oily food like chilly, Sichuan pepper, pepper, etc. should be reduced. That menu can nourish the liver and protect the spleen and stomach. It can not only ensure the health of the liver and abundant liver blood support, but also avoid the damage to the spleen and stomach caused by overeating fat, sweet and thick taste.

惊蛰美食 | FINE FOODS IN THE WAKING OF INSECTS

在我国民间素有“惊蛰吃梨”的食俗。“梨”者，“离”也，意为在害虫复苏之日即与害虫别离，以保一年里人体健康，庄稼不生虫害，五谷丰登。据明朝医学家李中梓的《本草通玄》记载，梨有“生者清六附之热，熟者滋五脏之阴”的保健功能。在惊蛰日食梨，无论生吃还是熟食，都是一种寓食于节的民俗传承。惊蛰时节，气温乍暖还寒，气候比较干燥，人们容易患上感冒、咳嗽等呼吸系统疾病，而梨有润肺清心、消痰降火的功效，吃梨极有益处。 此外，南方一些客家人还在惊蛰时食用煮熟、蘸糖的芋头，甜甜糯糯，美味又健脾。

In Chinese folk tradition, there is a custom of eating pear in the Waking of Insects. Pear in mandarin sounds like leaving. Leaving the pest when it wakes up will keep the body health, crops health and good harvest in the coming year. Li Zhongzi, medical scientist in Ming Dynasty (1368-1644), has written in his *Ben Cao Tong Xuan* that the pear has good healthcare function. It can clear away the waste heat inside our human body and moistening our organs. Eating fresh pear or cooked pear on the day of the Waking of Insects is an old tradition, in which food is an important part in this festival. In the Waking of Insects, the climate is dry, and the temperature rise and fall unexpected. It is easy to get respiratory diseases like influenza and cough. Pear has the function of cleaning lung and dissolving phlegm, hence eating pear is helpful. Moreover, some Hakkas in south China eat cooked taro with sugar in the Waking of Insects. It is sweet and soft, delicious and good for spleen.

椒麻春笋鸡

Jiaoma Flavor Chicken with Spring Bamboo Shoots

食材配方

春笋100克　熟鸡肉30克　红辣椒10克　葱青叶50克　花椒3克　鲜汤20毫升　食盐2克　酱油3毫升　味精1克　芝麻油10毫升

制作工艺

① 春笋入沸水焯水后撕成细丝；红辣椒切成细丝；熟鸡肉撕成细丝。

② 花椒用清水浸泡5分钟，捞出后与葱青叶一起剁成细末，加入鲜汤、食盐、酱油、味精、芝麻油调匀成椒麻味汁。
③ 将春笋丝装入盘中垫底，放上鸡丝，淋入椒麻味汁，点缀上红辣椒丝即成。

风味特色

色泽鲜艳明亮，鸡丝细嫩，春笋脆爽，椒麻味浓。

创意设计

惊蛰时节，乍暖还寒，常易外感风寒，宜多食新鲜蔬菜及蛋白质丰富的食材。此菜采用春季的代表性蔬菜——春笋与鸡肉搭配，调以椒麻汁，成菜具有高蛋白、低脂肪的特性，并含有大量膳食纤维，营养和食疗价值较高。此外，春笋有化痰之效，还能生发阳气，搭配健脾益胃的鸡肉，更能温中益气；而川菜特有的椒麻味可辛散温通，尤宜此时。

Ingredients

100g spring bamboo shoots; 30g pre-cooked chicken; 10g red chili peppers; 50g spring onion greens; 3g Sichuan peppercorns; 20ml stock; 2g salt; 3ml soy sauce; 1g MSG; 10ml sesame oil

Preparation

1. Blanch the spring bamboo shoots and cut into thin slivers. Cut the red chili peppers into thin slivers. Hand shred the precooked chicken into thin slivers.
2. Soak the Sichuan peppercorns for 5 minutes, remove and chop finely with the spring onion greens. Add the stock, salt, soy sauce, MSG and sesame oil, and blend well to make Jiaoma flavor sauce.
3. Spread the spring bamboo shoot slivers at the bottom of a plate, and top with the chicken slivers. Pour over the Jiaoma flavor sauce, and garnish with the red chili pepper slivers on the top.

Features

bright color; tender chicken; crispy spring bamboo shoots; strong Jiaoma flavor

Notes

At the time of the Waking of Insects, the weather is still cold though spring has begun. It is easy to catch a cold. It is suggested that people eat more fresh vegetables and foods rich in protein. Using chicken and spring bamboo shoots, a typical vegetable in spring, as the main ingredients and paired with Jiaoma flavor sauce, the dish is rich in protein and dietary fibers and low in fat with high nutritional and dietary therapy values. Among all the ingredients, spring bamboo shoots are able to reduce phlegm as well as invigorate yang and qi. Besides, chicken is able to tonify the spleen and nourish the stomach. With the two ingredients combined, the dish is mild in nature and benefits qi. Furthermore, the special Jiaoma flavor in Sichuan cuisine is best for the time of the Waking of Insects because of its spicy taste and body-warming function.

八宝釀梨

Steamed Pears Stuffed with Eight Treasures

食材配方

雪梨4个　莲子40克　糯米60克　薏仁40克　百合20克　枸杞6克　红枣10克　蜜樱桃5克
川贝2克　冰糖碎200克　清水100毫升

制作工艺

① 雪梨从蒂部用U型戳刀取下约2厘米大的一块作盖，掏去梨核及部分内瓤后去皮，入清水中浸泡备用。

② 糯米、百合、莲子、薏仁、红枣、川贝分别蒸熟，蜜樱桃切细，加入枸杞、冰糖碎拌匀，制成八宝馅料，放入梨中，加上梨盖，入笼蒸20分钟后取出装盘。

③ 锅置火上，加入清水和冰糖碎，用小火熬至浓稠后浇淋在雪梨上即成。

风味特色

成菜造型美观，味道甘甜爽口。

创意设计

在四川民间一向有“惊蛰吃梨”的食俗，以期远离害虫和疾病，祝愿人体健康。此菜选用雪梨与川贝等多种原料搭配，制法考究，口感丰富，味道甜美。在养生方面，雪梨、川贝能止咳化痰、清热降火，莲子健脾，糯米温中，薏仁利水，百合滋阴，枸杞养肾，红枣养心，辅以冰糖，清润之余，更能平调脏腑。

Ingredients

4 pears; 40g lotus seeds; 60g glutinous rice; 40g coix seeds; 20g lily bulbs; 6g wolfberries; 10g red dates; 5g candied cherries; 2g chuanbei (fritillaria cirrhosa); 200g rock sugar, crushed; 100ml water

Preparation

1. Cut off the top of each pear with U-shaped poking knife to make a 2cm cover. Core and peel the pears, and soak in water.
2. Steam the glutinous rice, lily bulbs, lotus seeds, coix seeds, red dates and chuanbei till cooked through. Chop the candied cherries, add the wolfberries and rock sugar, and blend well. Mix all the ingredients to make the eight-treasure filling. Transfer the filling into the pears and cover. Lay the pears in a steamer and steam for 20 minutes. Transfer to a serving dish.
3. Heat water and rock sugar in a wok over a low flame till the sugar melts and the soup becomes viscous and thick. Pour the soup over the pears.

Features

beautiful appearance; sweet, succulent and aromatic tastes

Notes

There is a custom in Sichuan that people eat pears at the time of the Waking of Insects for the purpose of keeping away from pests and diseases and wishing for a good health. This dish is of exquisite cooking method, delicate and sweet tastes. In food therapy, pears and chuanbei are able to relieve coughing, reduce sputum and clear up inner heat. Louts seeds help to tonify the spleen; glutinous rice is mild in nature; coix seeds promote body water excretion; lily bulbs help to nourish yin; wolfberries are able to nourish the kidney; and red dates help to tonify the heart. The rock sugar in the dish is able to clear heat and generate body fluid as well as nourish the internal organs.

芋头甜点 | Taro Dessert

食材配方

芋头90克　清水160毫升　籼米粉60克　葡萄干碎20克　砂糖10克　椰蓉50克　薄荷叶4朵　罐头红樱桃4个。

制作工艺

① 芋头切丁，放入破壁机中加入清水搅打成汁，先加入籼米粉搅匀，再加入葡萄干碎、砂糖搅匀制作成芋头糊。

② 将芋头糊倒入模具中至大约2厘米的高度，上锅蒸制15分钟至成熟后取出、放凉，再切成长约6厘米、宽与厚约2厘米的条，在表面沾裹满椰蓉后装盘，点缀上红樱桃、薄荷叶即成。

风味特色

成菜椰香浓郁，味道酸甜，质地软糯。

创意设计

惊蛰时节，南方一些客家人有食用甜芋头的习俗，以期生活甜美。四川一些地区的客家移民较多，也承袭了惊蛰食芋头的传统。此菜以芋头搅打成汁为主料制成，便是应此节气之品。在养生方面，芋头健脾补虚，所含水溶性多糖及黏液蛋白有抗氧化、减脂、降血糖、调节免疫力等作用，辅以砂糖、椰蓉、葡萄干，有强化益气补中之功，酸甜之味更符合此时节宜“省酸增甘”的饮食理念。

Ingredients

90g taro; 160ml water; 60g non-glutinous rice flour; 20g raisins, finely chopped; 10g sugar; 50g shredded coconut; 4 mint leaves; 4 canned red cherries

Preparation

1. Cut the taro into cubes and put into high speed blender. Add water and blend into juice. Add rice flour and blend well. Add raisins and sugar, and blend well to make taro paste.
2. Pour the paste in a mould to 2cm-high level and steam in a steamer for 15 minutes till cooked through. Leave to cool and cut into 6cm long, 2cm wide and 2cm high strips. Coat the strips evenly with shredded coconut and transfer to a serving dish. Garnish with mint leaves and red cherries.

Features

soft and glutinous texture; sweet and sour tastes with strong coconut flavor

Notes

At the time of the Waking of Insects, some Hakka people in the South have the custom of eating sweet taro in the hope of living a sweet life. There are lots of Hakka people in Sichuan, so the custom have been inherited. The main ingredient of this dish is taro, which is suitable for the Waking of Insects. In food therapy, taro helps to tonify the spleen. The water-soluble polysaccharide and mucinous proteins contained in taro are antioxidants and help to lose fat, regulate blood sugar level and immunity response. With sugar, shredded coconut and raisins added, the dish helps to tonify qi and the sour-sweet taste is accorded with the dietary principle of eating more sweet but less sour food in spring.

THE SPRING EQUINOX

仲春郊外

唐·王勃

东园垂柳径，
西堰落花津。
物色连三月，
风光绝四邻。
鸟飞村觉曙，
鱼戏水知春。
初晴山院里，
何处染嚣尘。

春分物候 | PHENOLOGY IN THE SPRING EQUINOX

春分是二十四节气中的第四个节气。当太阳位于黄经0° 时为春分，时间为每年的3月20日~22日。元朝吴澄《月令七十二候集解》言：“春分，二月中。分者，半也。此当九十日之半，故谓之分”“正阴阳适中，故昼夜无长短”。春分，古时又称“日中”“日夜分”“仲春之月”，一是古时以立春至立夏为春季，春分是春季九十天的中间点，平分了春季；二是指一天时间白天与黑夜平分，各有12小时。

The Spring Equinox is the fourth of the 24 Solar Terms. It begins when the sun reaches the celestial longitude of 0°. It usually begins from 20th to 22nd March every year. In his book *A Collective Interpretation of the Seventy-Two Phenological Terms*, Wu Cheng (Yuan Dynasty 1271-1368) said that the Spring Equinox is mid-February (clunar calendar); in Chinese we call it “Chun Fen”. “Fen” means half; day time is as long as night time. In ancient times, the Spring Equinox is called middle month of spring, etc. Firstly, ancient China regard the time from the Beginning of Spring to the Beginning of Summer as spring, and the Spring Equinox is the middle of spring and split spring fifty-fifty. Secondly, day time and night time are equal, 12 hours for each part.

我国古代将春分后的十五天又分为三候：一候元鸟至；二候雷乃发声；三候始电。“元鸟”即黑色的鸟，为燕子的别称，是春分来而秋分去之鸟，春分节气后，阳气始盛，气候温暖，秋归的燕子便从南方飞来了；雷，为阳气之声，由阴阳相交、相搏而产生雷声；“始电”，即因雷而发出霹雳闪电。春分相应的花信为：一候海棠；二候梨花；三候木兰。

Ancient China has divided the 15 days after the Spring Equinox into 3 pentads. In the first pentad, swallows will arrive. Swallow is suppose to come in the Spring Equinox and go in the Autumn Equinox; it is getting warm after the Spring Equinox, then swallow has come back from the south. In the second pentad, thunder will be heard. In the third pentad, flash will appear because of thunder. The news of flowers blooming in the Spring Equinox are as follows: begonia for the first pentad, pear blossom for the second pentad and magnolia for the third pentad.

食材生产 | PRODUCING FOOD INGREDIENTS

春分时节，我国大部分地区越冬作物都已进入春季生长阶段，正是春管、春耕、春种的关键时期，农谚有“春分麦起身，一刻值千金”“惊蛰早，清明迟，春分播种正当时”等说法。在资阳地区，当地有农谚说“惊蛰春分，红萝卜挂缨缨”。此时，四川盆地的气温回升较快，日均气温稳定上升到12℃以上，人们将抓住时机播种早稻、早玉米，育好红薯苗，适时浇灌小麦等，积极推进食材生产。

In the Spring Equinox, overwintering crops in most parts of China are in the spring growth stage. It is the critical time of spring managing, spring ploughing and spring sowing. As the farmer’s proverbs go, the Spring Equinox is very important for the growth of wheat; it is too early to sow in the Waking of Insects, too late to sow in Qingming, and it is the right time to do it in the Spring Equinox. In Ziyang region, there is a farmer’s proverb says that radish grow quickly in the Waking of Insects and the Spring Equinox. In this period, the

temperature is rising quickly in Sichuan Basin. The daily average temperature is above 12℃. People will take this opportunity to plant early rice and early corn, to breed sweet potato's seedling and irrigate wheat, in order to promote food ingredients production.

土地对于食材生产来说十分重要，长久以来，人们逐渐形成了对土地神即社神的自然崇拜。自汉代始，我国民间许多地方会在农历的二月、八月确定两个“社日”进行春秋社祭，即“春社”和“秋社”，时间一般是在立春和立秋后的第五个“戊”日，“春社”约在春分前后。据《荆楚岁时记》言：“社日，四邻并结综会社牲醪，为屋于树下，先祭神，然后飨其胙。”按唐宋旧俗，四川地区此时要“礼后土、演剧，乡村是日祭句芒神”。明清以后，在四川乡村有洗净耒耜、悬于梁上等习俗。这一天，农民不到田间劳作，而是将糯米捣成团，也有加入艾蒿者，燃香烛供之，谓“敬雀王”，即祭祀雀鸟，希望它们不食稼穑，四川民间也将其称为“敬春分馍馍”。

Land is very important for the production of food ingredients. Since long ago, people have started to worship god of the land. Since Han Dynasty(202 BC-220), people have fixed two days in February and August of lunar calendar to worship god of the land. Normally, the specific dates are the fifth Xu Day after the Beginning of Spring and the Beginning of Autumn. Spring celebration is just around the Spring Equinox. According to *Records of Festivals in Jingchu Region*, on that day people and their neighbors would gather together to worship the god under trees with sacrificing offerings, and then share all the offerings. According the old custom from Tang Dynasty and Song Dynasty, people would perform local opera to worship spring god in Sichuan region. After Ming Dynasty and Qing Dynasty, there is a custom of clearing plough and hanging it from a beam in Sichuan countryside. On the day of the Spring Equinox, farmers would stop working in fields. They would make dough with glutinous rice and Chinese mugwort. And then people would dedicate that dough, incense sticks and candles to king of birds. Farmers hope that birds would not eat their crops. That is called worship spring and share dough in folk tradition.

饮食养生 | DIETARY REGIMEN

古人云：“春分者，阴阳相半也，故昼夜均而寒暑平。”我国传统医学和养生学认为，春分时节正是调理人体内部阴阳平衡、协调机体功能很好的养生时机，此时需抑制冬季旺盛的阴气，提升略弱的阳气，调整偏颇体质。春分时节，四川盆地虽然气温回升较快，但由于冷空气活动频繁，天气多变，有时会出现“倒春寒”，因此，在饮食养生上，首先要“以平为期”，保持阴阳寒热的均衡。可根据个人体质和食物属性进行合理搭配，如烹调鸭肉、河蚌、河蟹等寒性食物时，最好配合散寒的葱、姜、黄酒等；食用韭菜、韭黄、蒜苗等热性食物时，最好配合养阴的蛋类、猪肉等，如韭菜炒鸡蛋、蒜苗回锅肉等。其次要多食甘味食物，可补养脾胃、健脾祛湿，如谷米、山药、鸡肉、鸭肉等即可补养脾胃，也可选用薏米、茯苓、赤小豆等与鲫鱼、鲤鱼同炖，以健脾祛湿。再次，可适当吃一些“春菜”，以应和自然界的春季与人体肝脏生发的要求，如应季的香椿、豆芽、蒜苗、豆苗、莴苣、韭菜、菠菜等春菜。另外，春分时节切忌偏颇饮食，少食性味过于偏寒、偏热或升降浮沉太过的食物。

The ancients said, the Spring Equinox means yin and yang are equal; day time and night time are equal, it is not cold and not hot. TCM believes that the Spring Equinox is the proper time to balance yin and yang in human body, and adjust the body functions. In this period, the yin qi developed in winter should be restrained, and the weak yang qi should be encouraged. In the Spring Equinox, although the temperature in Sihcuan Basin increases quickly, late spring coldness happens sometime because of the frequent cold air activities. Therefore, dietary regime then should obey the normal standard and keep the balance. People can design diet according to individual physique and the natures of foods. For example, when we cook "cold" foods like duck, freshwater mussel, river crab, it is better to add green onion, ginger and yellow rice wine which can clear coldness. When we eat "hot" foods like leek, chives, garlic bolt, it is better to add eggs and pork, like Scrambled Egg with Leek, Double Cooked Pork Slices with Garlic Bolt, etc. In this period, people need eat foods in "sweet" feature, like rice, Chinese yam, chicken, duck which can tonify spleen and stomach. Crucian and carp could be cooked with pearl barley, poria, red bean, which can nourish spleen and clear humidity. Considering the connection between spring and operation of liver, people can eat some spring vegetables, like seasonal toona sinensis, bean sprout, garlic bolt, bean seedling, lettuce, leek, spinach, etc. In the Spring Equinox, people should avoid biased diet. We should eat less foods with too much "cold" and "hot" feature.

春分美食 | FINE FOODS IN THE SPRING EQUINOX

春分时节，过去四川农家有吃“菜卷子”的习俗，现在川西部分地区仍有保留。菜卷子的制作方法为：以过半糯米、小半大米混匀，泡涨后磨浆，滴干多余水分后切为小块，内包菜肉馅，外裹菜叶，用笼或甑蒸熟即食，也可晾凉后储存，待食时再蒸热即可。俗语道：“春日宴，啖水鲜。”冬日里潜藏的鱼虾随着春雷的一声声呼唤，从水底上游至水面，开始享用鲜嫩的新生水草和众多的浮游生物，其肉质更为鲜美。因此，每年的3月底至4月初是品尝河鲜的最佳时节。

In the Spring Equinox, Sichuan farmers have the custom of eating vegetable rolls, which has been kept nowadays in west Sichuan. The way of cooking vegetable rolls are as follows: mix more glutinous rice and less rice together, grind them into thick liquid after soaking in water, get dough from the liquid, put mince inside and vegetable outside, steam it right now or dry it for further cooking in future. As the old saying goes, it is the proper time to eat river food in spring. Fishes and shrimps have swum to the water surface to get fresh foods after a long winter. Hence, they are delicious too. It is the best time to eat river food at the turn of March and Arpil.

此外，由于春社的时间与春分相近，而春社是立春后的第五个戊日，按五行来说，戊属土，为土神忌日，古人认为在这一天下地干活常会出现庄稼多虫害的现象，于是，包括四川在内的许多地区，农家都会把各种农作物的果实，如红豆、扁豆、黄豆、玉米等放在锅里混炒后让家人同食，这一习俗被称作“炒虫”“吃虫”，以祈求减少农作物的病虫害。

Moreover, the date of Spring Celebration (we call "Chun She" in Chinese) is the fifth Xu Day after the Beginning of Spring, close to the Spring Equinox. According to the theory of five primary elements, Xu belongs to earth element and is the day of God of land. Ancient people believed that working in fields on that day will cause more pests attacks. Therefore, in lots of places including Sichuan, farmers will fry fruits of many crops together, like red bean, lentil, soy bean, corn, and eat them with family members. It is called fry pests, and is a way of praying for less pests attack.

蒜香小河鱼

Garlic Flavor Stone Moroko

食材配方

小河鱼200克　大蒜50克　小葱30克　洋葱30克　胡萝卜20克　青椒20克　红椒20克　食盐3克　料酒10毫升　蒜末50克　干辣椒10克　干花椒3克　香菜20克　鸡蛋60克　淀粉50克　食用油1500毫升（约耗60毫升）

制作工艺

① 大蒜、小葱、洋葱、胡萝卜、青椒、红椒、香菜、食盐、料酒入碗，制成腌料汁水；鸡蛋与淀粉调成全蛋淀粉糊；蒜末入锅炸香。

② 锅置中小火上，入食用油烧至90℃时，加入干辣椒节、干花椒炒香后取出，用刀铡成双椒末。

③ 小河鱼制净，放入腌料汁水中腌制15分钟，捞出后晾干水分。

④ 锅置火上，入食用油烧至160℃时，将小河鱼用全蛋淀粉糊拌匀后放入锅中炸，至色泽金黄、质地酥脆后捞出装盘，配上双椒末、蒜末、香菜味碟即成。

风味特色

河鱼色泽金黄，质地外酥里嫩，味道丰富，最宜佐酒。

创意设计

资阳境内的沱江鱼类品种丰富，小河鱼产量较大，因其刺多，故最宜油炸。此菜用蔬菜汁腌渍河鱼，更能去腥增鲜，炸制后更细嫩、香浓。在养生方面，鱼肉利水健脾却性偏凉，搭配温热的蒜、葱、辣椒等，能互为中和，比较适宜此时节“平和”的饮食需要。

Ingredients

200g stone moroko; 50g garlic; 30g spring onions; 30g onions; 20g carrots; 20g green chili peppers; 20g red chili peppers; 3g salt; 10ml Shaoxing cooking wine; 50g finely chopped garlic; 10g dried chilies; 3g Sichuan peppercorns; 20g coriander; 60g eggs; 50g cornstarch; 1,500ml cooking oil (60ml to be consumed); 50g starch

Preparation

1. Combine the garlic, spring onions, onions, carrots, green chili peppers, red chili peppers, coriander, salt and Shaoxing cooking wine in a bowl, and mix well to make the marinade. Mix the egg and starch well to make egg paste. Stir fry the finely chopped garlic till aromatic.
2. Heat up oil to 90℃ in a wok over a medium-low flame, add dried chili segments and dried Sichuan peppercorns, and stir fry till aromatic. Finely chop the Sichuan peppercorns and dried chilies.
3. Rinse the stone moroko and marinate for 15 minutes. Remove and drain.
4. Heat up the oil to 160℃ in a wok, coat the stone moroko evenly with egg paste and deep fry till golden brown and crispy. Transfer to a serving dish. Mix the Sichuan peppercorns, dried chilies, fried garlic and coriander to make the dipping sauce. Serve with the fish.

Features

golden brown color; tender inside but crispy on the outside; suitable as an accompaniment to baijiu (Chinese liquor)

Notes

The Tuojiang River within the Ziyang City of Sichuan boasts abundant fish species, among which stone moroko is easy to be caught. However, stone moroko has a lot of bones, and that is why it is often deep fried. The fish is marinated in vegetable juice to remove the fishy smell and enhance the fresh taste. After deep frying, it is more tender and aromatic in taste. In food therapy, fish helps to tonify the spleen, but it is cold in nature. With garlic, spring onions and chilies that are warm in nature, the dish is mild on the whole, suitable for the Spring Equinox.

什锦炒豆

Assorted Fried Beans

食材配方

干胡豆50克　干黄豆50克　干豌豆50克　小米辣椒20克　青尖辣椒15克　椿芽10克　藿香10克　姜米10克　蒜米15克　葱花10克　酱油10毫升　芝麻油5毫升　醋10毫升　食盐10克　生抽10毫升　白糖20克　凉开水100毫升

制作工艺

① 小米辣椒、青尖辣椒切成长约0.5厘米的颗粒；椿芽、藿香分别切细。

② 将小米辣椒、青尖辣椒、姜米、蒜米、葱花、酱油、醋、食盐、生抽、芝麻油、白糖、凉开水入碗调成味汁。

③ 锅置中小火上，入干胡豆、干黄豆、干豌豆烘炒至熟，起锅放入味汁中浸泡2小时，至其入味后捞出装碗即成。

风味特色

成菜质地酥脆，味道咸鲜微辣，乡土气息浓郁。

创意设计

春分时节不仅常与春社相近，且多在农历二月中，四川许多地方在此时有吃炒豆的习俗，以期减少病虫害。此菜将三种豆类炒熟后再用调味料浸渍入味，既是习俗的传承，又改善了口感，适宜性更强。在养生方面，胡豆、豌豆、黄豆三者同用，能和中下气、利水解毒，且富含蛋白质及膳食纤维，可降低“三高”，延缓衰老，但却有胀气之弊，故搭配藿香、椿芽，可化湿行气、避免胀闷。

Ingredients

50g dried broad beans; 50g dried soybeans; 50g dried peas; 20g bird's eye chilies; 15g green chili peppers; 10g Chinese toon sprouts; 10g huoxiang (agastache rugosus); 10g ginger, finely chopped; 15g garlic, finely chopped; 10g spring onions, finely chopped; 10ml soy sauce; 5ml sesame oil; 10ml vinegar; 10g salt; 10ml light soy sauce; 20g sugar; 100ml cold boiled water

Preparation

1. Slice the bird's eye chilies and green chili peppers into 0.5cm granules. Shred Chinese toon sprout and huoxiang.
2. Make the seasoning source with the bird's eye chilies, green chili peppers, ginger, garlic; spring onions, soy sauce, vinegar, salt, light soy sauce, sesame oil, sugar and cold boiled water.
3. Heat a wok over medium-low flame, and roast the dried broad beans, dried soybeans and dried peas till cooked through. Soak in the seasoning sauce for two hours, and transfer to a serving bowl.

Taste

crispy texture; salty delicate, and slightly spicy tastes; a typical rural dish

Notes

The the Spring Equinox usually comes in the second lunar month, and is close to Chunshe, a traditional spring ceremony worshiping the earth god in the hope of a good harvest. Around the Spring Equinox, people in and around Sichuan have the custom of eating fried beans to keep healthy and protect themselves from injurious insects. The beans are roasted and soaked in specially made sauce. This dish is not only an inheritance of local culinary tradition but also improves the taste of beans. The dish is also great for people's health. Broad beans, soybeans and peas regulate qi in human body, induce water excretion and promote body detoxication. They are also rich in protein and dietary fibers, have anti-aging effects, and help reduce blood pressure, blood fat, and blood glucose. With agastache rugosus and Chinese toon sprouts, this dish relieves flatulence and clears dampness.

春分菜卷子

the Spring Equinox Vegetable Rolls

食材配方

糯米粉400克　籼米粉100克　猪油20克　温水400毫升　猪绞肉（肥4瘦6）200克　碎米芽菜100克　冬笋50克　料酒7毫升　胡椒粉0.5克　食盐0.5克　酱油10毫升　姜末3克　味精1克　白糖5克　花椒粉1克　芝麻油3毫升　葱花20克　生菜400克　食用油30毫升

制作工艺

① 将糯米粉、籼米粉、猪油用温水调制成软硬适中的米粉面团。

② 冬笋洗净，切成小丁；锅置火上，加入食用油烧热，下猪绞肉炒散，之后加入食盐、酱油、姜末、料酒、胡椒粉和味精炒香，再放入冬笋丁炒匀后出锅，加花椒粉、芝麻油和葱花拌匀制成馅心。

③ 面团搓条、下剂，包入馅心，捏成长筒形，放入笼屉后蒸熟，取出卷上生菜，切掉两端即成。

风味特色

成菜皮白软糯，馅心咸鲜微麻。

创意设计

菜卷子是春分时节四川一些地区人们的应节之品。此菜在制作时添加冬笋丁为馅心，并以新鲜生菜叶取代传统的树叶来卷包，营养价值与可食性更高。此外，猪肉、生菜、冬笋性凉，糯米性温，一凉一温，一阴一阳，共同搭配，可平调阴阳，而且冬笋富含有维生素、矿物质及多糖，能促进肠道蠕动，有助于消化。

Ingredients

400g glutinous rice flour; 100g non-glutinous rice flour; 20g lard; 400ml warm water; 200g pork mince (40% lard and 60% lean); 100g yacai (preserved mustard stems); 50g winter bamboo shoots; 7ml Shaoxing cooking wine; 0.5g pepper; 0.5g salt; 10ml soy sauce; 3g ginger, finely chopped; 1g MSG; 5g sugar; 1g ground Sichuan pepper; 3ml sesame oil; 20g spring onions, finely chopped; 400g lettuce; 30ml cooking oil

Preparation

1. Make rice dough with glutinous rice flour, non-glutinous rice flour, lard and warm water.
2. Rinse the winter bamboo shoots, and cut into small cubes. Heat cooking oil in a wok, and stir fry the pork mince. Add the salt, soy sauce, ginger, cooking wine, pepper and MSG, and stir till fragrant. Blend in the winter bamboo shoots, stir well and remove from the wok. Mix well with ground Sichuan pepper, sesame oil and spring onions for stuffing.
3. Knead the dough into strips, cut into small lumps, and press and flatten. Wrap up the stuffing with the dough, and shape into long cylinders. Steam in a steamer till cooked through, and remove. Wrap up the cylinder with lettuce, and cut off the two ends.

Features

white and glutinous wrappers; savory and slightly numbing stuffing

Notes

Vegetable Rolls is one of people's favorite dishes on the Spring Equinox in some places in Sichuan Province. This dish has higher nutritional value as winter bamboo shoots are used as stuffing and edible fresh lettuce instead of tree leaves is used as skin. Pork, lettuce and winter bamboo shoots are cold in nature while glutinous rice warm in nature. The opposite natures of the ingredients make the nutrition more balanced. Particularly, winter bamboo shoots improve intestinal peristalsis and digestion with the various vitamins, minerals and polysaccharide contained.

QINGMING

清明
气清景明 杨柳风轻
清明
唐·杜牧
清明时节雨纷纷，
路上行人欲断魂。
借问酒家何处有，
牧童遥指杏花村。

清明物候 | PHENOLOGY IN QINGMING

清明是二十四节气中的第五个节气。当太阳到达黄经15° 时为清明，时间在每年的4月4日~6日。元朝吴澄《月令七十二候集解》言："清明，物至此时皆以洁齐而清明矣。" 汉朝《历书》说："盖时当气清景明，万物皆显，因此得名。" 意思是说，此时天气和万物清新明朗。

Qingming (or Pure Brightness) is the fifth of the 24 Solar Terms. It begins when the sun reaches the celestial longitude of 15°. It usually begins from 4th to 6th April. In his book *A Collective Interpretation of the Seventy-Two Phenological Terms*, Wu Cheng (Yuan Dynasty 1271-1368) said that everything is bright and clear in Qingming. Another book *Almanac* written in Han Dynasty said that the climate and all things are clear and bright in this period.

我国古代将清明后的十五日分为三候：一候桐始华； 二候田鼠化为鴽；三候虹始见。清明五日后，白桐木的花首先绽放；再过五日，喜阴的田鼠躲入洞穴，属阳的鹌鹑等小鸟开始活跃；又经五日，雨后的天空可开始出现彩虹。清明相应的花信为：一候桐花；二候麦花；三候柳花。

Ancient China has divided the 15 days after Qingming into 3 pentads: in the first pentad, the white paulownia blossoms; in the second pentad, shade-loving voles hide in caves and sunshine-loving quails become active; in the third pentad, rainbow appears after rain. The news of flowers blooming in Qingming are as follows: paulownia blossom for the first pentad, wheat blossom for the second pentad and willow blossom for the third pentad.

食材生产 | PRODUCING FOOD INGREDIENTS

清明一到，四川地区的气温显著升高，雨水稍多。农谚有"清明前后种瓜种豆" "清明时节，麦长三节"等，催促人们及时进行食材种植和田间管理。清明时节的天气仍然变化不定，忽冷忽热，时阴时晴，对已萌芽和返青生长的农作物危害很大，因此，人们常常还需在此时进行防寒、防冻工作。清明之后，资阳地区的胡豆、豌豆、樱桃等食材开始成熟上市。

In Qingming, the temperature of Sichuan region rises remarkably and the rain fall is a little more. As the farmer's proverbs go, "plant melons and beans around Qingming", and "wheat will grow 3 sections longer in Qingming". Those proverbs urge people to plant food ingredients and manage the fields. The climate in Qingming is not stable yet. It is fluctuant between cold and hot, cloudy and sunny, which is very harmful for crops that are sprouting and turning green. Therefore, people normally need do cold-proofing and freeze-proofing work then. After Qingming, lima bean, pea, cherry are in the market in Ziyang region.

四川地区历来非常重视水利，远在战国时期，蜀郡太守李冰就主持修建了闻名中外的大型水利工程——都江堰，"旱则引水浸润，雨则杜塞水门"，使整个川西平原"水旱从人，不知饥馑，时无荒年"。因此，每年清明节，一年一度的清明"放水节"是四川农业生产的重要习俗之一。

Sichuan has pay great attention to hydra-engineering since long ago. In Warring States Period, Li Bing, chief of Shu county, has built the world famous Dujiangyan Irrigation System. "Let water in during drought and block the dam when raining", which has made the west Sichuan plain a real place of abundance. Hence, there is a very important Water Pouring festival in every Qingming.

饮食养生 | DIETARY REGIMEN

清明时节，自然界阳气渐盛，天气虽然日渐暖和，但昼夜温差依然较大，且晴雨多变。此时，雨水较多、湿气较重，易使人疲倦嗜睡，也容易感冒，应少吃辛热、过于生发的食物，如羊肉、狗肉、花椒、辣椒、胡椒、白酒等，可多吃补脾利湿的食物，如山药、扁豆、糯米、薏苡仁等。同时，由于阳气已生发，人体内阴气已很微弱，而肝气至此达到最旺，需防止上火，尤其是预防肝火旺：一方面，肝气应发挥其疏泄、调节气血的重要作用，宜适当多吃补肝、补血的食物使肝血充沛，如樱桃、枸杞子、核桃、花生、红枣等；另一方面，肝气过于强盛则易产生肝郁和肝火而伤脾、肺，所以应食用一些疏肝、清肝、健脾、润肺的食物，如菠菜、油菜、草莓、金橘等新鲜蔬果。此外，可用玫瑰花、金橘及绿色时蔬等制作菜点，也可吃一些时令野菜，如苦菜、马兰、荠菜等，以缓解内热及春季干燥引起的出鼻血等症。同时，白菊花茶能平肝明目，绿茶能清利肝火，金银花茶清热疏风，山药、扁豆、谷米、蜂蜜能益气健脾，均可根据需要适量食用。

In Qingming, yang qi is getting stronger. Though temperature is rising, the temperature difference between day and night is still large, and it is fluctuant between sunny and rainy. It is raining more in this period and the humidity will make people feel tired and drowsy, and will easily catch cold. People should eat less pungent and "hot" food, like mutton, Sichuan pepper, chilly, pepper, white liquor, etc. It is better to eat more foods which can tonify spleen and clear humid, like Chinese yam, lentil, glutinous rice and coix seed, ect. Meanwhile, yin qi inside human body is very weak and energy of liver is the strongest in this period, we must prevent excessive internal heat and protect liver. On the one hand, we must make liver play the right role by eating food which can tonify liver and enrich the blood, like cherry, boxthorn, walnut, peanut, red date. On the other hand, in order to prevent strong liver energy affecting spleen and lung, we should take some foods which can soothe liver and tonify spleen and moisten lung, like spinach, rape, strawberry, kumquat, etc. Moreover, we can make dishes with rose, kumquat and green seasonal vegetable; and eat some seasonal potherb like sow thistle, kalimeris indica, leaft mustard, to prevent nosebleed because of spring dryness. Furthermore, white chrysanthemum tea can clam liver and benefit eye; green tea can relieve liver heat; honeysuckle tea can clear the internal heat; Chinese yam, lentil, grain and honey can tonify our energy and spleen when taking proper amount.

清明美食 | FINE FOODS IN QINGMING

四川地区的清明习俗与其他地区相似，祭祖与踏青、饮食并行。清代嘉庆《资阳县志》云："三月清明，男女具牲肴祭于墓，挂纸幡拜泣，是日民人多具酒肴出城游玩聚饮，名曰踏青。"此外，还有放风筝、荡秋千、吃清明糕等习俗。四川民间清明扫墓敬祖时，常有举办"清明会"的习俗，合族凑钱买酒，家族聚餐，吃"清明糕"。清明糕是采摘田野里的清明草（又称鼠曲

草、棉花草）拌以糯米粉捣揉成团，包上甜豆沙或白萝卜、春笋、腊肉丁等馅蒸制而成，色泽青碧，有止咳化痰的作用。也有用艾叶等制成的清明糕或清明果，又称为“艾蒿馍馍”。

Qingming customs in Sichuan are similar to other regions. There are customs of worshiping ancestors, spring outing and special foods. *Ziyang Local Chronicles* in Jiaqing, Qing Dynasty recorded that, "in Qingming, people worship the ancestors in front of the tombs. People play outside the city with liquor and dishes, which is called spring outing." In addition to that, there are customs of flying kites, swinging and eating Qingming cake. When people worship ancestors in Sichuan, they would hold a Qingming party. Relatives would pool money to buy dishes and liquors, and eat Qingming cake. People make Qingming cake with Qingming grass, glutinous rice flour, sweet bean paste, radish, spring bamboo shoot, preserved pork particles. The steamed cake can relieve cough and reduce sputum. Some people will make Qingming cake with mugwort leafs, such cake can also be called mugwort steamed bread.

随着时代的发展，清明节又糅合了古代寒食节等文化内涵。寒食节常常是在清明前一天，古代有禁火、吃“馓子”等食俗。馓子为油炸面点小吃，味道香脆，古时称为“寒具”。如今，禁火习俗在我国大部分地区已不流行，但馓子依旧深受大众喜爱。

With the development of the times, Qingming Festival has covered the culture connotation of ancient Cold Food Festival. Cold Food Festival is just one day before Qingming. There were customs of fire prohibition and eating Sanzi. Sanzi is a kind of fried pastry snacks, delicious and crispy, which was called cold food in the old days. Nowadays, fire prohibition is not accepted in most parts of China, but Sanzi is still popular.

樱桃肉 | Cherry Pork

食材配方

带皮五花肉200克　鲜樱桃100克　生姜10克
大葱15克　料酒15毫升　糖色30克　高汤500毫升
白糖30克　食盐3克　麦芽糖20克　食用油30毫升

制作工艺

① 带皮五花肉切成约1厘米大小的正方形，入沸水中，加入生姜、大葱、麦芽糖煮至6成熟时捞出，入油锅中炸成金黄色。

② 鲜樱桃洗净，一半放入榨汁机榨成樱桃汁。

③ 锅置火上，放入高汤、五花肉、料酒、糖色、白糖、食盐、鲜樱桃、樱桃汁烧沸，改用小火烧至肉质软熟，色泽红亮，汤汁浓稠后起锅装盘，点缀上樱桃即成。

风味特色

成菜色泽红亮，味道酸甜，肉质软糯。

创意设计

樱桃肉是一道广受大众喜爱的传统川菜，但是因为味道过甜而不太符合现代健康饮食要求。此菜在原来的作法上加以改进，采用二次烧制法去掉部分脂肪，并加入鲜樱桃烧制，甜酸适中，鲜樱桃的香浓渗透在肉中，老少皆宜。在养生方面，樱桃调中补气、益肾健脾、生津止渴，为时令佳果，与猪肉搭配能清热滋阴。此外，樱桃含铁量居水果之首，胡萝卜素丰富，能调节睡眠、清除自由基、抗癌、抗氧化。

Ingredients

200g pork belly; 100g fresh cherries; 10g ginger; 15g spring onions; 15ml Shaoxing cooking wine; 30g caramel; 500ml stock; 30g sugar; 3g salt; 20g malt sugar; 30ml cooking oil

Preparation

1. Cut the pork belly into 1cm cubes, add to boiling water, and blend in the ginger, green onions and malt sugar. Boil till the pork is medium well, remove, and deep fry till golden.
2. Rinse the cherries, and feed half of them in a juicer for juice.
3. Place a wok over the flame, add the stock, pork belly, cooking wine, caramel, sugar, salt, fresh cherries and cherry juice, and bring to a boil. Turn down the heat and continue to braise over a low flame till the pork is soft and reddish brown and the stock reduced. Transfer to a serving dish and garnish with cherries.

Features

reddish brown in color; sweet and sour tastes; soft and glutinous texture

Notes

Traditional Cherry Pork is a popular Sichuan dish but people nowadays lose passion for it as it tastes too sweet and does not meet the demand for healthy diet. In view of this, two-time heating is taken to get rid of extra fat in the pork, and fresh cherry juice gives the dish an appetizing sweet and sour taste. So the improved version is favored by the old and young. It also has great health benefits. Cherries strengthen qi in human body, tonify the kidney and spleen, produce saliva and slake thirst. Coupled with pork, cherries clear inner heat and nourish yin. In addition, cherry has the highest iron rate among all fruits and rich carotene, which improves people's sleeping, scavenges free radicals, and prevents cancer and oxidation.

翡翠豆花

食材配方

鲜蚕豆100克　黄豆100克　酥黄豆15克　馓子20克　大头菜颗20克　芹菜颗20克　红油辣椒20毫升　酱油10毫升　醋10毫升　味精1克　花椒粉1克　香菜5克　葱花10克　葡萄糖酸内酯3克

制作工艺

① 鲜蚕豆去皮、洗净；黄豆用清水浸泡至透，先将两豆搅打成豆浆，再加入葡萄糖酸内酯水溶液搅匀，倒入盆中上笼蒸5分钟使其凝固成豆花后取出。

② 将豆花装入碗中，依次加入酥黄豆、馓子、大头菜颗、芹菜颗、红油辣椒、酱油、醋、味精、花椒粉、香菜、葱花即成。

风味特色

成菜色泽淡绿，质地细嫩，味道咸鲜，略带麻辣。

创意设计

豆花是传统川菜的代表品种之一，从素豆花到荤豆花，其食材、做法和吃法各不相同。此菜采用资阳出产的鲜蚕豆、黄豆等制作而成，色泽清雅，淡绿似翠，故而得名。养生方面，豆花有清热解毒、生津润燥之效，在此季节食用，正当其时，可防止肝火过旺、清热降火；此外，鲜蚕豆与黄豆共用，二者互补，更有益于提升蛋白质的营养价值。

Ingredients

100g fresh broad beans; 100g soybeans; 15g crispy soybeans; 20g fried dough twists; 20g preserved kohlrabi, finely chopped; 20g celery, finely chopped; 20ml chili oil; 10ml soy sauce; 10ml vinegar; 1g MSG; 1g ground Sichuan pepper; 5g coriander; 10g spring onions, finely chopped; 3g delta-gluconolactone

Preparation

1. Peel and clean the fresh broad beans, and soak the soybeans in water. Mix the beans and beat to make bean milk. Add delta-gluconolactone to the milk, blend well, pour into a basin, and steam for 5 minutes for doufu pudding.
2. Remove the pudding to several bowls, and add crispy soybeans, fried dough twists, kohlrabi, celery, chili oil, soy sauce, vinegar, MSG, ground Sichuan pepper, coriander and spring onions.

Features

verdant color; tender doufu pudding; savory and spicy tastes

Notes

Doufu pudding is one of the representative Sichuan dishes. However, ingredients used and the ways of preparing and eating may vary from place to place. Some people prefer vegetarian doufu pudding while others like meat toppings. Jade Doufu Pudding is made from fresh broad beans and soybeans grown in Ziyang City. It is virid and looks like jade, hence it's named Jade Doufu Pupdding. It is also great for people's health. Doufu pudding helps clear heat of liver especially in this season, remove toxicity and produce saliva. Combining fresh broad beans and soybeans, this dish provides more balanced nutrition and high quality protein.

清明粿 | Qingming Cake

食材配方

糯米粉250　籼米粉25克　澄粉20克　艾蒿叶250克　洗沙馅250克　肉松50克

制作工艺

① 艾蒿叶洗净，放入沸水中焯熟后取出，用冷水漂冷后榨成艾蒿汁。

② 澄粉加入沸水烫熟，与糯米粉、籼米粉混合均匀，加入艾蒿汁调制成软硬适中的米粉面团。

③ 将米粉面团下成剂子，包上洗沙馅和肉松，收紧封口，再捏成圆球状，分别放入铝盏中成生坯。

④ 将做好的生坯放入蒸笼中，用旺火蒸约8分钟至熟后取出装盘即成。

风味特色

外皮色泽碧绿，口感质地软糯，馅心香甜爽口。

创意设计

清明时节以野菜制作面点小吃，是四川及全国很多地方的传统习俗之一，品种众多。此点以糯米粉和艾蒿汁调成粉团，包上洗沙馅和肉松制成，色绿质软，香甜清爽，营养和食疗价值较高。其中，艾叶性温，可祛湿散寒，且有抗菌抑菌作用，与糯米搭配食用，能有效防止脾虚湿困及疲倦嗜睡等现象；豆沙性寒，能辅助祛湿，寒热中和，防止上火。

Ingredients

250g glutinous rice flour; 25g non-glutinous rice flour; 20g gluten-free wheat flour; 250g mugwort leaves; 250 mashed red beans; 50g pork floss

Preparation

1. Rinse the mugwort leaves, blanch in boiling water, remove and leave to cool in cold water. Process the mugwort leaves in a blender to make mugwort juice.
2. Add boiling water to the gluten-free wheat flour, and mix well with glutinous rice flour and non-glutinous rice flour. Add the mugwort juice, knead and roll to make dough.
3. Cut the dough into small lumps, wrap up the mashed red beans and pork floss, and seal. Shape into round balls, and place into the aluminum cups.
4. Steam the balls in a steamer over a high flame for 8 minutes till cooked through, remove and transfer to a serving dish.

Features

green and glutinous skin; sweet and refreshing stuffing

Notes

Around Qingming Festival, it is a custom of Chinese people to eat cakes or snacks made from wild vegetables, and people in different places make different Qingming foods. It has considerable nutrition and health values. Mugwort leaves are warm in nature, and help to clear inner dampness, dispel inner cold and prevent bacteria growth. When used with glutinous rice, the herb prevents spleen dampness and drowsiness. Red beans, cold in nature, clear up body dampness and neutralizes the heating effects of mugwort.

GRAIN RAIN

谷雨

布谷啼播
雨洗纤素

谷雨

清·郑板桥

不风不雨正晴和，
翠竹亭亭好节柯。
最爱晚凉佳客至，
一壶新茗泡松萝。
几枝新叶萧萧竹，
数笔横皴淡淡山。
正好清明连谷雨，
一杯香茗坐其间。

谷雨物候 | PHENOLOGY IN GRAIN RAIN

谷雨是二十四节气中的第六个节气。当太阳到达黄经30°时为谷雨，时间为每年的4月19日~21日。元朝吴澄《月令七十二候集解》言："谷雨，自雨水后，土膏脉动，今又雨其谷于水也。"意思是此时节下雨，会滋润百谷茁壮成长，因此，谷雨寓"有雨百谷生"之意。

Grain Rain is the sixth of the 24 Solar Terms. It begins when the sun reaches the celestial longitude of 30°. It usually begins from 19th to 21st April. In his book *A Collective Interpretation of the Seventy-Two Phenological Terms*, Wu Cheng (Yuan Dynasty 1271-1368) said that rain fall in this solar term would moisten all the grains and promote their growth. Hence, Grain Rain means that all the grains grow when there is rain.

我国古代将谷雨后的十五日分为三候：一候萍始生；二候鸣鸠拂其羽；三候戴胜降于桑。谷雨后降雨增多，浮萍开始生长；布谷鸟不住地抖动羽毛，按捺不住地放声歌唱；桑树上也开始见到戴胜鸟了。谷雨相应的花信是：一候牡丹；二候荼蘼；三候楝花。

Ancient China has divided the 15 days after Grain Rain into 3 pentads: in the first pentad, duckweed begins to grow; in the second pentad, cuckoo loudly sings, while shaking its feathers; in the third pentad, hoopoe appears on mulberry. The news of flowers blooming in Grain Rain are as follows: peony for the first pentad, roseleaf raspberry for the second pentad and Chinaberry flower for the third pentad.

食材生产 | PRODUCING FOOD INGREDIENTS

我国长江流域有"清明下种，谷雨下秧"的农谚，资阳地区的农谚则言"清明谷雨，小春雄起"，鼓励人们鼓足干劲进行农业生产。在四川盆地，谷雨前后的降雨常常"随风潜入夜，润物细无声"，这是因为"巴山夜雨"以四五月份出现得最多。这种夜雨昼晴天气对大春作物生长和小春作物收获是颇为有宜的。

In Yangtze River Basin, there is a farmer's proverb, "seed in Qingming; transplant rice seedling in Grain Rain". In Ziyang region, farmer's proverb has encouraged people to work hard in fields in Qingming and Grain Rain. In Sichuan Basin, the rainfall around Grain Rain usually happens in evening quietly. It is mostly raining during the night of April and May. Such kind of weather is very helpful for the growth of crops sown in spring and crops harvested in the next year.

俗语说"清明见芽，谷雨见茶"。到谷雨时节，随着气温升高，茶树的芽叶生长加快，积累的内含物也更为丰富，因此雨前茶往往滋味鲜浓而耐泡。四川民间也在此时采茶、制茶、交易茶。同时，在谷雨前后，资阳地区的桑葚逐渐成熟，一年一度的桑葚采摘节吸引了大量的游客前往尝新。

As the proverb goes, "the tea sprouts appear in Qingming, and the tea leaves grow in Grain Rain". In Grain Rain, as temperature rises, tea leaves grow quickly and get more nutriment inside. Tea before Grain Rain tastes fresh and delicious, and good to brew. Sichuan people pick tea leaves, make tea and trade tea in this period. Meanwhile, mulberry is ripe around Grain Rain. The Annual Mulberry Picking Festival has attracted many tourists.

饮食养生 | DIETARY REGIMEN

四川地区的气候因为暮春谷雨时节的来临而致阳气渐盛、天气温热，4月下旬平均气温已接近20℃，但因多雨之故，空气湿度较大；而春阳内应于肝，又易引起肝阳上亢。所以，谷雨时节在饮食养生方面，可适当食用一些具有养血柔肝、补血益气，以及清热降火、祛湿润燥的甘平、甘凉食物，如动物肝脏、河鲜及湖鲜等鱼类食物，可养血柔肝；此外，白扁豆、赤小豆、薏苡仁、山药、冬瓜、白萝卜等食物具有良好的清热祛湿作用，也可适当食用一些；还可用草菇等菌类食材与豆腐制羹，以补血益气。值得注意的是，谷雨时节不宜进食羊肉、狗肉、麻辣火锅，以及辣椒、花椒、胡椒等大辛及大热之品，白酒也要少饮，以防助火升阳。养生谚语说“谷雨夏未到，冷饮莫先行”，因此，在这个时节要少吃冷食，以免挫伤阳气，损伤脾胃。同时要注意饮水，尤其是晨起喝一杯温开水，银耳、桑葚、蜂蜜能润燥养阴，可适量食用。此外，味甘微苦、性凉的绿茶有清热生津之效，也可酌量泡饮。

In Sichuan region, it is getting warm and yang qi has become strong because of the arrival of Grain Rain and late spring. The average temperature in the second half of April is close to 20℃. And it is humid because of more rain fall. Yang qi in spring is correspondent to liver, and the energy of liver will rise then. Hence, in Grain Rain, people should eat some foods with sweet and neutral nature, which can nourish blood, tonify liver, clear internal heat, moisten dryness and dispel dampness, like animal livers, river fishes and lake fishes. Moreover, people can eat white hyacinth bean, red bean, coix seed, Chinese yam, winter melon, white radish, which have good functions of clearing internal heat and dispel dampness. People can also make soup with straw mushroom and doufu, which can nourish blood and promote energy. It is worth noting that, people should not eat mutton, hot pot, chilly, Sichuan pepper, pepper and drink less white liquor in Grain Rain. Those food ingredients will promote the internal heat. As the proverb goes, "Don't take cold drink in Grain Rain, because the summer has not come." Hence, people should eat less cold foods, which are harmful to stomach in this solar term. Meanwhile, drinking water especially drinking warm water in morning is very important. People can take proper amount of tremella, mulberry, honey to moisten dryness. Green tea with bitter taste and cold nature can clear the internal heat, and good to drink.

谷雨美食 | FINE FOODS IN GRAIN RAIN

明代许次纾在《茶疏》中谈到采茶时节时说："清明太早，立夏太迟，谷雨前后，其时适。"谷雨茶是指谷雨时节采制的春茶，又叫二春茶。到了谷雨前后，四川地区气温适中、雨水充沛，加上茶树经过冬季的休养生息而芽叶肥嫩、色泽翠绿、叶质柔软，富含多种维生素和氨基酸。用谷雨这天上午采摘的茶叶做成的干茶，才算得上是真正的谷雨茶。它除了嫩芽外，还有一芽一嫩叶或一芽两嫩叶的特点，与清明茶即头春茶相似，同为茶中佳品。中国茶叶学会等有关部门曾提出倡议，主张将每年农历谷雨这一天作为"全民饮茶日"。

Xu Cishu from Ming Dynasty has said in his *Book of Tea* (written in 1597) that "it is too early to pick tea in Qingming; it is too late to pick tea in the Beginning of Summer; it is just the right time around Grain Rain." Grain Rain Tea is the spring tea made in Grain Rain, which is called second spring tea too. Around Grain Rain in Sichuan region, the temperature is pleasant, and the rain is abundant. The young leaves of the tea trees, which have rested for a long winter, is green, soft and full of all kinds of vitamin and amino acid. Only the tea picked in the morning of the day of Grain Rain can be called Grain Rain Tea. That tea has the formation of one sprout plus one young leaves, or one sprout plus two young leaves. Just like Qingming tea which is also called first spring tea, it is among the best teas. China Tea Science Society has proposed to make the day of Grain Rain the national day of drinking tea.

谷雨前后也是牡丹花盛开的重要时段，因此，牡丹花也被称为"谷雨花"。"谷雨三朝看牡丹"，赏牡丹已成为人们暮春闲暇时的重要娱乐活动之一，四川彭州等地在谷雨时节都会举办牡丹花会，供人游赏。此外，牡丹花可食用，我国从宋代开始就兴起食用牡丹花的习惯，到明清时，人们已经研发出很多跟牡丹花有关的食谱，据清代顾仲的《养小录》记载："牡丹花瓣，汤焯可，蜜浸可，肉汁烩亦可。"

It is also the important period for peony to blossom around Grain Rain. Hence, people call peony Grain Rain Flower. Appreciating peony in Grain Rain is one of the important relaxations in late spring. Pengzhou area in Sichuan will host peony fair in Grain Rain. Furthermore, peony is edible. Since Song Dynasty, Chinese people have the habit of eating peony. In Ming and Qing Dynasty, people have invented many menus for peony. As the book *Yang Xiao Lu* written by Gu Zhong of Qing Dynasty recorded, peony petals can be eaten with soup, honey and meat gravy.

茶香虾 | Tea Shrimps

食材配方

小河虾100克　桑叶茶30克　薄荷2克　椒盐6克
薄荷尖1朵　食用油500毫升（约耗50毫升）

制作工艺

①桑叶茶用沸水泡涨，捞出沥干水分，入油锅中炸至酥脆。

②小河虾洗净，剪掉虾须，加入食盐和茶水浸泡入味，捞出沥干水分，入油锅中炸至酥香后捞出。

③将桑叶茶入盘中垫底，放上小河虾，撒上椒盐，点缀上薄荷尖即成。

风味特色

河虾外酥里嫩，桑叶茶脆香爽口，最宜佐酒。

创意设计

桑叶茶是资阳特产。此菜以河虾与桑叶茶为食材制成，是一道具有浓郁资阳特色、又有益健康的美食。桑叶疏风散热、平肝明目，薄荷清利头目、疏肝行气，与补肾益精的河虾搭配，在增加特殊风味之时，又寓补于食。此外，河虾肉质鲜嫩，高蛋白、低脂肪；桑叶茶富含氨基酸、维生素及多种生理活性物质，有降血压、降血糖、延缓衰老等作用。

Ingredients

100g river shrimps; 30g mulberry tea leaves; 2g mint; 6g Sichuan pepper salt; 1 mint; 500ml cooking oil (about 5ml to be consumed)

Preparation

1. Pour boiling water over the mulberry tea leaves, and leave to soak. Remove, drain, and deep fry till crispy. Save the tea water.
2. Rinse the river shrimps, and cut off their whiskers. Add salt and tea water, and leave to marinate. Remove, drain, and deep fry till crispy and aromatic.
3. Spread the mulberry leaves on a serving dish, and place the shrimps on top. Sprinkle with pepper salt and garnish with the mint.

Features

shrimps tender inside but crispy on the outside; a strong mulberry leave aroma; a good accompaniment to baijiu (Chinese liquor)

Innovation

Mulberry leaf tea is a local specialty of Ziyang. Using mulberry tea leaves and shrimps as the main ingredients, this healthy dish is characteristic of Ziyang. In terms of food therapy, mulberry leaves help to remove inner heat, boost liver function and benefit the eye; mint clears up the eye and the head, relieves the liver and promotes the circulation of qi inside the body; and river shrimps are beneficial for the kidney. Besides, shrimps are tender in texture, high in protein but low in fat. Mulberry tea, which contains rich amino acids, vitamins and physiologically active substances, lowers blood pressure and glucose, and retards the aging process.

牡丹鱼片

Peony Fish Slices

食材配方

花鲢鱼净肉500克　胡萝卜20克　小黄瓜100克　鲜香菇30克　青笋10克　巧克力酱20克
红油30毫升　白糖20克　食盐2克　花椒油3毫升　土豆泥30克　食用油1000毫升（约耗30毫升）

制作工艺

① 将花鲢鱼肉切成厚约0.1厘米的薄片后晾干；胡萝卜雕成花心。
② 锅置火上，入食用油烧至100℃时，放入鱼片炸成金黄色后捞出，加入红油、白糖、食盐、花椒油拌匀。
③ 小黄瓜制成树叶形状；青笋做成花枝。
④ 将鱼片插在土豆泥上制成牡丹花形，再做好花叶、枝干，将制好的香菇放在底部，用巧克力酱写上字即成。

风味特色

造型美观，刀工精细，色泽红亮，质地酥脆，味道咸鲜，略带麻辣。

创意设计

牡丹花被誉为“国花”，因常在谷雨时节盛开，故又名“谷雨花”。此菜采用传统川菜灯影牛肉的做法，将鱼片切成薄片后炸制成花瓣，再拼摆出牡丹花形，既体现了川菜技艺的精湛，也形似于现实中的牡丹花。养生方面，花鲢鱼肉温中益气、利水消肿，高蛋白、低脂肪，食疗和营养价值突出。

Ingredients

500g silver carp meat; 20g carrot; 100g cucumbers; 30g shitake mushrooms; 10g Chinese lettuce stems; 20g chocolate sauce; 30ml chili oil; 20g sugar; 2g salt; 3ml Sichuan pepper oil; 30g mashed potatoes; 1,000ml cooking oil (about 30ml to be consumed)

Preparation

1. Cut the fish into 0.1cm slices and drain. Cut and use the carrot as the center of a flower.
2. Heat the cooking oil in a wok to 100℃, and deep fry the fish slices till golden brown. Remove, add chili oil, sugar, salt and Sichuan pepper oil, and blend well.
3. Cut the cucumbers into the shape of leaves, and the lettuce stems the shape of a flower stem.
4. Insert the fish slices into mashed potatoes in the shape of a peony. Put in place the leaves and stem. Lay in the shitake mushrooms. Write with chocolate sauce.

Features

vivid presentation; exquisite cutting techniques; bright colors; savory and slightly spicy tastes; crispy texture

Innovation

Peony, known as the national flower, is also called grain rain flowers as they often blossom during the solar term of Grain Rain. This dish draws on the cooking methods of a traditional Sichuan dish, Shadow Beef, and reflects the fine craftsmanship of Sichuan cuisine. In food therapy, silver carp helps to promote qi and relieve edema. Besides, fish is nutritious with high protein and low fat.

桑叶锅摊

Mulberry Leaf Pancakes

食材配方

面粉150克　食用桑叶50克　清水350毫升　鸡蛋2个　鲜牛奶50毫升　熟白芝麻5克　食盐8克　玉米淀粉30克　白糖3克　食用油5毫升

制作工艺

① 桑叶洗净，加入清水50毫升榨成桑叶汁水。
② 面粉和玉米淀粉混合均匀，加入鸡蛋液、鲜牛奶、食盐、白糖、桑叶汁、熟白芝麻和清水300毫升搅匀呈稀面糊。
③ 平底锅置火上，入食用油烧热，倒入稀面糊摊成薄饼，煎熟后起锅装盘即成。

风味特色

色泽深绿，质地软糯，味道咸甜，桑叶香浓。

创意设计

桑叶是乐至县的特色食材。此点以新鲜食用桑叶汁为食材制成，具有桑叶独特的清香。养生方面，桑叶是“药食同源”食材，可平肝明目、疏风散热，能避免春季后期肝阳上亢；面粉益气和中，二者搭配，有平肝健脾之效。此外，桑叶能抑制脂肪肝、动脉粥样硬化，还有降血糖、降血脂、降胆固醇及抑菌、抗衰等作用。

Ingredients

150g wheat flour; 50g edible mulberry leaves; 350ml water; 2 eggs; 50ml milk; 5g roasted white sesames; 8g salt; 30g cornstarch; 3g sugar; 5ml cooking oil

Preparation

1. Rinse the mulberry leaves, add water (about 50ml), and process in a blender till smooth.
2. Mix the wheat flour with cornstarch, add the eggs, milk, salt, sugar, mulberry leaf juice, sesames and water (about 300ml), and stir well.
3. Heat oil in a pan, add the mixture, and bake till cooked through. Transfer the pancakes to a serving dish.

Features

verdant in color; soft in texture; salty and sweet in taste with a strong mulberry leaf fragrance

Innovation

Mulberry leaves are often used as ingredients in Lezhi County. In food therapy, mulberry leaves, which are both food and herbal medicine, help to boost the liver function, benefit the eye and remove inner heat which is quite common in late spring. Wheat flour relieves the liver and kidney. Besides, mulberry leaves prevent fatty liver disease and atherosclerosis, and lower blood sugar, blood fat and cholesterol. They are also antibacterial and anti-aging.

THE BEGINNING OF SUMMER

立夏
气序清和 万物并秀
阮郎归·初夏
宋·苏轼
绿槐高柳咽新蝉，
薰风初入弦。
碧纱窗下水沈烟，
棋声惊昼眠。
微雨过，小荷翻，
榴花开欲然。
玉盆纤手弄清泉，
琼珠碎却圆。

立夏物候 | PHENOLOGY IN THE BEGINNING OF SUMMER

立夏是二十四节气中的第七个节气，也是夏季的第一个节气。当太阳到达黄经45° 时为立夏，时间为每年的5月5日～7日。战国末年，立夏在我国就已经确立。元朝吴澄《月令七十二候集解》言："立夏，四月节。立字见解春。夏，假也。物至此时皆假大也。"意思是说各种植物到此时已经直立长大，故名立夏。它标志着春天的结束、夏季的来临，因此，古人又称立夏为"春尽日"。

The Beginning of Summer is the seventh of the 24 Solar Terms, and the first solar term in summer. It begins when the sun reaches the celestial longitude of 45°. It usually begins from 5th to 7th May every year. The concept of the Beginning of Summer have been fixed at the end of Warring States Period in China. In his book *A Collective Interpretation of the Seventy-Two Phenological Terms*, Wu Cheng (Yuan Dynasty 1271-1368) said that all kinds plants have grow up in this solar term, hence we call it the Beginning of Summer. It marks the end of Spring and the arrival of summer. Ancient Chinese also call the Beginning of Summer ending day of spring.

中国古代将立夏后的十五天分为三候：一候蝼蝈鸣；二候蚯蚓出；三候王瓜生。立夏五日后，蝼蛄（即蝼蝈，俗名拉拉蛄、蝲蝲蛄、土狗等）开始在田间鸣叫；再过五日，地下的蚯蚓开始翻松泥土；再经五日，王瓜的蔓藤开始快速攀爬生长。

Ancient China has divided the 15 days after the Beginning of Summer into 3 pentads: in the first pentad, mole crickets begin to chirp in the fields; in the second pentad, earthworms start to dig underground; in the third pentad, the vines of trichosanthes cucumeroides grow fast.

食材生产 | PRODUCING FOOD INGREDIENTS

立夏时节，万物进入快速生长期，也是农业生产和食材生长的一个关键期。据古文献记载，周朝时，帝王亲率文武百官在立夏的这一天到郊外"迎夏"，并命令官员到各地督促百姓抓紧农业耕作。立夏日的农谚较多，如"立夏看夏"，意为冬小麦扬花灌浆、夏收作物收成几近定局；"多插立夏秧，谷子收满仓"，此时插秧的多少，直接关系到秋天的收获；"立夏不下，犁耙高挂""立夏无雨，碓头无米"，而此时的劳作与雨水多少有关，并直接影响到日后的收成。

In the Beginning of Summer, all things on earth have entered the rapid growth period, and it is also a key period for agriculture and food ingredients producing. According to ancient historical record, emperors in Zhou Dynasty (1046 BC-256 BC) would welcome the summer in the suburbs with all his courtiers on the day of the Beginning of Summer. They would order officials to urge the agricultural production all over the country. There are many farmer's proverbs for the day of the Beginning of Summer, like "you can touch the summer when it comes" which means winter wheat has grown into the final stage and the harvest in summer is almost guaranteed; "plant more summer rice seedlings, fill warehouse with rice then", the number of summer rice seedlings will decide the harvest in autumn; "if there is no rain in the Beginning of Summer, there is no need to work in fileds", and "if there is no rain in the Beginning of Summer, there is no hope to get the rice in harvest time", which mean rain will decide amount of labor and the future harvest.

这一时节，除了辛勤耕种外，各地都有许多当季食材开始收获。在四川地区，子实丰富的油菜开始收割，人们压榨出新鲜的菜子油，制作成明亮的红油，用以烹制川菜；橙黄的枇杷已挂满枝头，资阳市雁江区的伍隍枇杷是其中的名品；乐至桑葚已经上市，颗粒饱满、色泽紫亮，甘甜无比。

In this solar term, people begin to harvest lots of seasonable food ingredients all over the country, in addition to hard work in fields. In Sichuan region, local farmers harvest well-grown oilseed rape, crush oil from the oilseed, and make bright chili oil to cook Sichuan cuisine. Orange colored loquat clings to the branches. The Wu Huang loquat from Yanjiang District, Ziyang City belongs to the first class. Mulberry from Lezhi County is on the market too. The bright purple mulberry's particle is satiate and very sweet.

饮食养生 | DIETARY REGIMEN

立夏时节，气温开始升高，阳气渐盛，气血向外运行，心脏负担加重。我国传统医学认为，心在五行之中属“火”，为夏季主时之脏，和暑气相通，主管人的神志。由于夏季暑热之气易亢而成邪，容易引发心脏功能失常、心烦失眠、心神不安等症状，因此，立夏要注意养心，可多选择番茄、赤小豆等红色食物。同时，立夏以后，四川地区逐渐炎热，雨水逐渐增多，容易引起人们食欲不振、胃口不佳，湿热病邪时有发生，因此，可选择冬瓜、丝瓜、苦瓜、杨梅、甜瓜、桃、李等蔬果，不仅应时而且能够解暑止渴。也可以选择制作时令粥品，如绿豆粥能清热解毒、荷叶粥能去浊醒脾，茯苓粥、竹叶粥能养心安神。

In the Beginning of Summer, the temperature is getting higher and higher. Growing yang qi brings more burden to the heart when this energy and blood circulate. TCM believes that the heart belongs to the fire primary element, and is the ruling internal organ in summer which is connected with summer heat and control human's mind. The summer heat is so strong to make the heart malfunctional, to cause unease and insomnia. Therefore, people need to protect heart in the Beginning of Summer by eating more red food like tomato, red bean, etc. After the Beginning of Summer, it is getting hot and rainy in Sichuan region. People will loose appetite and get illness caused by damp and hot. In order to relieve summer heat and thirst, we need eat winter

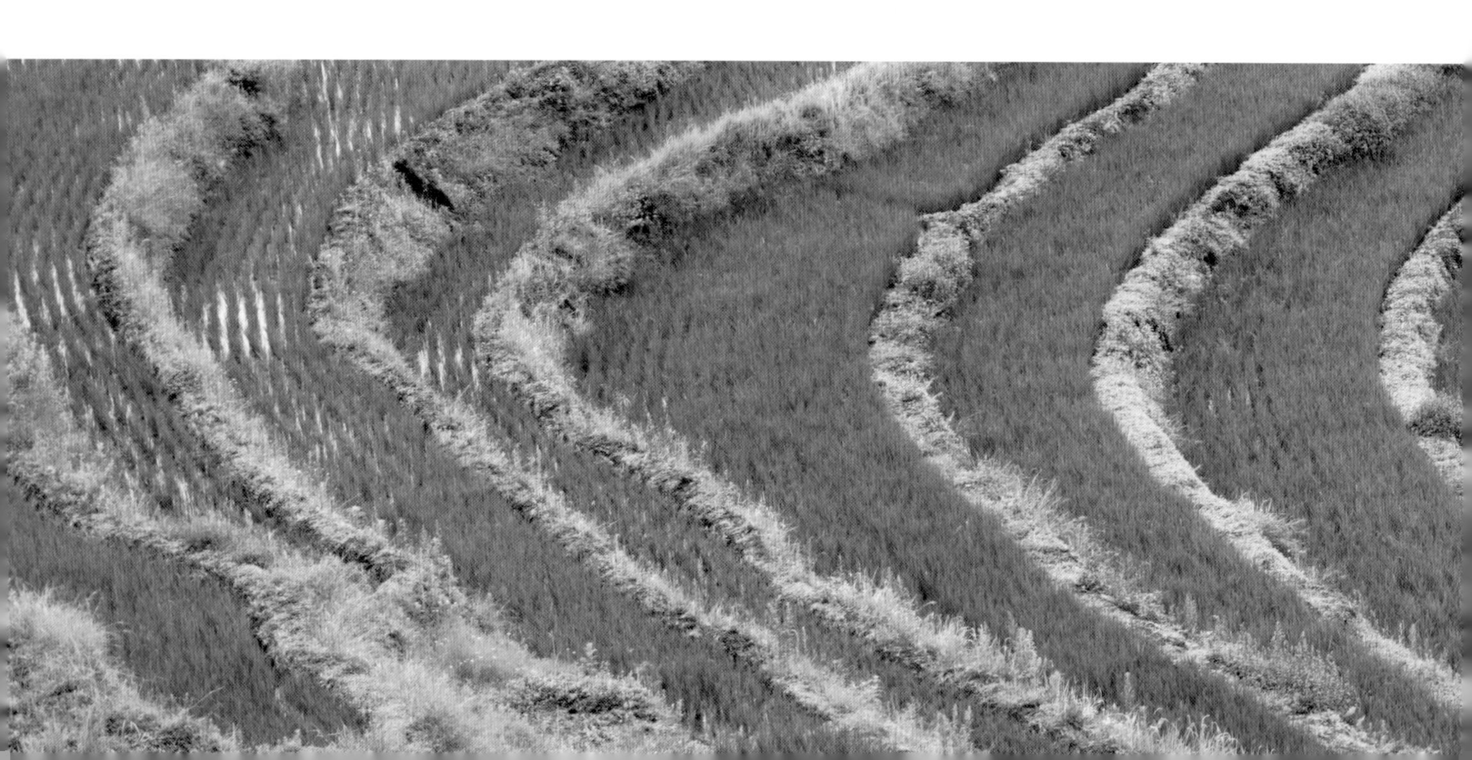

melon, towel gourd, bitter gourd, waxberry, muskmelon, peach, plum, etc. We can also make seasonable porridge. Porridge with mungbean can clear heat and remove toxicity; porridge with lotus leaves can remove the turbid and invigorate spleen; porridge with tuckahoe and porridge with bamboo leaves can tranquillize the mind by nourishing the heart.

立夏美食 | FINE FOODS IN THE BEGINNING OF SUMMER

立夏时节，在南方许多地区有“尝新”的食俗，包括吃五色饭（乌米饭）、立夏蛋和尝三鲜等。立夏当天，人们用绿豆、黄豆、青豆、黑豆、红豆等与大米混合煮制成“五色饭”，后来逐渐演变成蚕豆肉煮米饭，加上苋菜黄鱼羹，称之为“立夏饭”。在四川许多地区，立夏饭有豌豆糯米饭、煮鸡蛋、成对笋和带壳豌豆，民间有立夏吃鸡蛋拄（意为“支撑”）心、吃笋拄腿、吃豌豆拄眼之说。古谚云：“立夏吃了蛋，热天不疰夏”“立夏胸挂蛋，孩子不疰夏”。疰夏主要是指人们食欲减退、四肢无力、身体消瘦。吃鸡蛋能够增进食欲、增强体力，预防疰夏疾病。同时，民间还有“豆蛋”的习俗，即孩子们三三两两在一起相互碰蛋，以蛋壳坚而不碎者为赢。人们认为，立夏日吃蛋、斗蛋能祈祷夏日平安，经受疰夏的考验。

In the Beginning of Summer, there is a food custom called Chang Xin (have a taste of what is just in season), including eating Five-Colored Rice, the Beginning of Summer Egg and three delicacies, etc. On the day of the Beginning of Summer, people in south China make Five-Colored Rice by mung bean, soybean, green bean, black bean, red bean and rice. Later, people cook rice with broad bean and meat, and eat the rice with amaranth yellow croaker soup. That is the Beginning of Summer Rice. In lots parts of Sichuan, the Beginning of Summer Rice includes pea glutinous rice, boiled egg, bamboo shoots in pairs and peas with shell. People believe that eating egg in the Beginning of Summer is good for our heart; eating bamboo shoots can strengthen legs; eating pea can protect eyes. As the proverb goes, eating eggs in the Beginning of Summer will avoid the loss of appetite, body weakness and loss of weight. Moreover, there is a custom, that children break one egg with another and the unbreakable will be the winner. People believe that eating eggs and play this custom in the Beginning of Summer will bless the peace in summer.

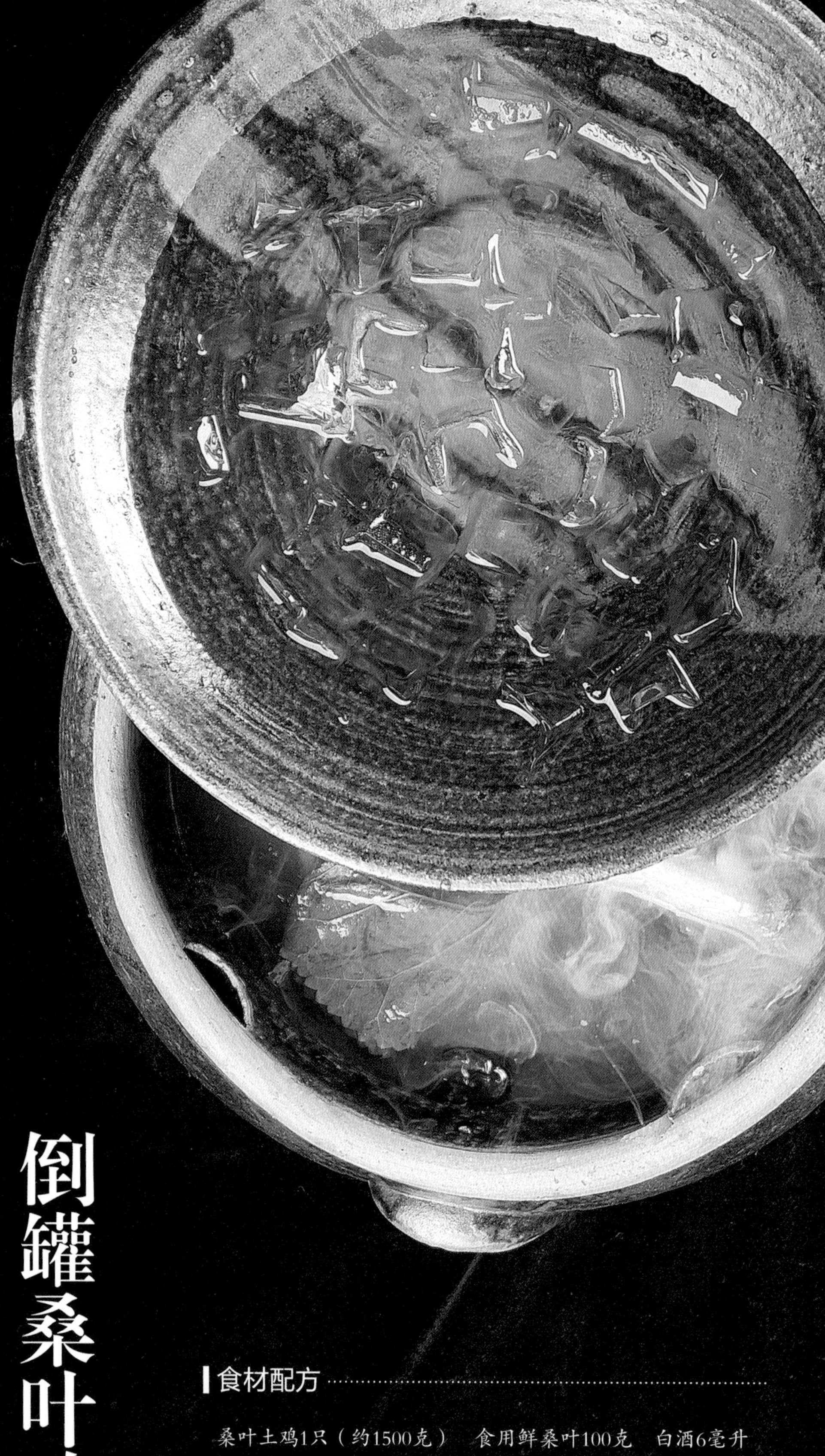

倒罐桑叶鸡

Steamed Chicken with Mulberry Leaves in a Reversion Crock

食材配方

桑叶土鸡1只（约1500克） 食用鲜桑叶100克 白酒6毫升

姜片15克 葱段30克 干辣椒节3克 花椒1.5克 胡椒粉1克

食盐10克 大枣3个 枸杞10克

制作工艺

①将整鸡抹上白酒搓揉10分钟，再抹上食盐、胡椒粉腌渍40分钟。

②把腌渍好的整鸡放入倒罐中，再依次加入鲜桑叶、姜片、葱段、干辣椒节、花椒、大枣、枸杞，盖上罐盖，并在盖子内注满冰水。

③将倒罐置于铁锅上，隔水蒸制6小时（在此期间需不断更换盖子内的冰水，让水温始终保持在20℃以内）至鸡肉软熟，最后将倒罐取出即成。

风味特色

成菜汤汁清澈，鸡肉嫩滑，味道咸鲜，桑叶香味浓郁。

创意设计

倒罐蒸土鸡是一道在乐至县回澜镇民间传承两百余年的传统名菜，蒸制时不加一滴水，全靠蒸汽倒流形成汤汁而得名。此菜选取用桑叶喂养的土鸡加上桑叶等食材蒸制而成，不但有桑叶特殊的香味，而且颇具食疗价值。其中，鸡肉温中益气、补精填髓，尤宜夏季温阳，蛋白质含量高，与桑叶同蒸，更利于益气养肝。

Ingredients

1,500g free-range chicken fed with mulberry leaves; 100g edible mulberry leaves; 6ml baijiu (a strong Chinese liquor); 15g ginger, sliced; 30g spring onions, segmented; 3g dried chilies, segmented; 1.5g Sichuan peppercorns; 1g pepper; 10g salt; 3 red dates; 10g wolfberries

Preparation

1. Smear the whole chicken skin with liquor and rub for 10 minutes. Smear salt and pepper on the skin and marinate for 40 minutes.

2. Place the chicken in a reversion crock, and add mulberry leaves, ginger, spring onions, dried chilies, Sichuan peppercorns, dates and wolfberries. Cover the crock with the cap, and fill the cap with ice water.

3. Put the crock in a wok and steam for 6 hours (keep changing the ice water in the cap in the process to ensure the temperature of the water is below 20℃) till the chicken is cooked through. Remove the crock from the wok.

Features

clear soup; tender chicken; savory taste; strong mulberry leaves aroma

Notes

Steamed Free-Range Chicken in a Reversion Crock is a traditional dish in Huilan Town of Lezhi County with a history of over 200 years. During the steaming, no water is added to the crock, and the chicken is cooked only by the vapor. With mulberry leaves and local free-range chicken fed with mulberry leaves as the main ingredients, this dish has unique mulberry aromas and considerable health benefits. Chicken can warm the stomach and spleen, replenish qi, an ideal food to dispel inner cold in summer. Steamed with mulberry leaves, high-protein chicken has better functions in promoting qi circulation and nourishing the liver.

姜汁鲜桑叶

Mulberry Leaves in Ginger Juice

食材配方

食用鲜桑叶20张　姜末15克　食盐4克　味精2克　鸡精5克　酱油2毫升　醋12毫升　芝麻油5毫升　鲜汤30毫升

制作工艺

①鲜桑叶入沸水锅中焯水后捞出晾凉、装盘。

②将姜末、食盐、味精、鸡精、酱油、醋、芝麻油、鲜汤入碗调成姜汁味，淋在桑叶上即可。

风味特色

色泽碧绿，质地细嫩，味道咸酸清香，姜味浓郁。

创意设计

立夏是采摘桑叶、桑葚的重要时节，而食用桑叶也是此时的重要食材之一。此菜将桑叶焯水后凉拌制成，最大限度保持了桑叶的清香。在养生方面，桑叶为药食两用食材，可疏散风热、清肺润燥、清肝明目，且有降血糖、降血脂、降血压、抗氧化等作用，佐以姜末，能避免凉性太过，利于夏季养阳、温肺开胃。

Ingredients

20 pieces of edible mulberry leaves; 15g ginger, finely chopped; 4g salt; 2g MSG; 5g chicken essence; 2ml soy sauce; 12ml vinegar; 5ml sesame oil; 30ml stock

Preparation

1. Blanch the mulberry leaves in boiling water, remove and leave to cool. Transfer to a serving dish.

2. Blend the ginger, salt, MSG, chicken essence, soy sauce, vinegar, sesame oil, stock in a bowl to make ginger-flavored seasoning, then drizzle over the mulberry leaves.

Features

pleasant green color; tender texture; fresh, savory and sour tastes; strong ginger flavor

Notes

It is time to pick mulberry leaves and mulberries during the Beginning of Summer. That's why edible mulberry leaf is a very important ingredient at this time. This dish preserves the fresh fragrance of mulberry leaves to the utmost extent by simply blanching them. In food therapy, mulberry leaves can disperse wind and purge heat, clear the lung and moisturize dryness, relieve the liver and improve vision. Pungent ginger used in this dish helps to balance the cold nature of mulberry leaves, warming the lung, increasing the appetite, and cultivating yang in summer. Besides, mulberry leaves contain antioxidant and also has the function of lowering blood sugar, blood pressure and cholesterol.

食材配方

胡萝卜200克　澄粉250克　猪油20克　面粉100克　鲜枇杷200克　白糖50克　可可粉10克　鹰粟粉（玉米淀粉）20克　黄油50克　熟花生仁20克

制作工艺

① 胡萝卜去皮、切薄块，入锅焯水断生后起锅漂冷，榨成胡萝卜汁，倒入汤锅中烧沸，加入澄粉烫成全熟面团后起锅，加入猪油揉匀，制成橘色面团；面粉与可可粉混合均匀，加入黄油40克和清水调制成褐色面团。

② 枇杷去皮，取果肉加工成泥状，加入白糖、鹰粟粉和黄油10克拌匀，放入蒸箱中蒸制成熟，制成枇杷馅。

③ 橘色面团搓成长条、下剂，包入枇杷馅心，捏成椭圆形，用褐色面团搓成细条、装饰成枇杷形，即成枇杷果生坯。

④ 将枇杷果生坯入笼蒸制8分钟至成熟，取出装盘即成。

风味特色

色泽橙黄，形似枇杷，质地软糯，味道甜香。

创意设计

立夏正是枇杷上市的最佳季节。此点以新鲜枇杷果肉为馅，以胡萝卜汁和澄粉等为皮制成枇杷果，作为立夏的象形面点。在养生方面，当季鲜果枇杷生津止渴、和胃降逆的效用极佳，与澄粉、面粉一起制作面点，更能益气养胃。此外，枇杷含有的三萜类化合物有抗炎、抗病毒、抗肿瘤和调节免疫力等作用。

Ingredients

200g carrots; 250g gluten-free wheat flour; 20g lard; 100g wheat flour; 200g loquats; 50g sugar; 10g cocoa powder; 20g cornstarch; 50g butter; 20g precooked peanuts

preparation

1. Peel the carrots and slice, then blanch till just cooked. Remove, cool down in water, and process in a blender to make carrot juice. Pour the juice in a pot, bring to a boil, and add the gluten-free wheat flour to make cooked dough. Add the lard and knead well to have orange dough. Mix wheat flour with cocoa powder, add 40g butter and appropriate amount of water to make a brown dough.
2. Peel the loquats, deseed and process into puree. Blend well with sugar, cornstarch and 10g butter. Steam in a steamer till cooked through to make the loquat stuffing.

枇杷果
Steamed Loquat Cake

3. Roll the orange dough into a strip and divide it into equal portions. Press and flatten into wrappers, put the loquat stuffing on wrappers and wrap up into oval shape. Roll the brown dough into thin strip, then garnish the orange dough pieces like a real loquat.
4. Steam loquat dough in a steamer for 8 minutes till cooked through, then remove and transfer to a serving dish.

Features

vivid loquat shape; golden orange in color; soft and glutinous texture; sweet and aromatic taste

Notes

The Beginning of Summer is the best time for loquats. This loquat cake adopts fresh loquats as the stuffing and carrot juice and wheat flour as the skin. It is a vivid symbol that summer is coming. As for food therapy, seasonal fresh loquats have great functions of producing saliva to quench thirst and relieving the stomach. Cooking fresh loquats with wheat flour, this snack has a better effect on replenishing qi and nourishing the stomach. Furthermore, loquat's triterpenes have anti-virus, anti-inflammatory, anti-tumor and immunity-enhancing effects.

小满

宋·欧阳修

夜莺啼绿柳，
皓月醒长空。
最爱垄头麦，
迎风笑落红。

LESSER FULLNESS OF GRAIN

小满
麦穗初齐
桑叶正肥

小满物候 | PHENOLOGY IN LESSER FULLNESS OF GRAIN

小满是二十四节气中的第八个节气。当太阳到达黄经60° 时为小满，时间为每年的5月20日～22日。元朝吴澄《月令七十二候集解》言："四月中，小满者，物致于此小得盈满。"意思是说，从小满开始，大麦、冬小麦等夏收作物籽粒渐渐饱满，但尚未成熟，处于乳熟后期。

Lesser Fullness of Grain is the eighth of the 24 Solar Terms. It begins when the sun reaches the celestial longitude of 60°. It usually begins from 20th to 22nd May every year. In his book *A Collective Interpretation of the Seventy-Two Phenological Terms*, Wu Cheng (Yuan Dynasty 1271-1368) said that since Lesser Fullness of Grain, barley and winter wheat are in late milk stage, and have grown well.

我国古代将小满后的十五天分为三候：一候苦菜秀；二候靡草死；三候麦秋至。小满五日后，苦菜变得枝繁叶茂，可以采摘食用；再过五日，各种杂草在阳光曝晒下开始枯死；再经五日，麦子成熟收获。从小满到芒种期间，降雨不断增多，南北温差进一步缩小。

Ancient China has divided the 15 days after Lesser Fullness of Grain into 3 pentads: in the first pentad, sowthistle has grown up and is good to eat; in the second pentad, weeds begin to wither in strong sunshine; in the third pentad, wheat can be harvested. From Lesser Fullness of Grain to Grain in Beard, rain fall is increasing and the temperature difference between south and north is narrowed further.

食材生产 | PRODUCING FOOD INGREDIENTS

小满时节，包括四川在内的许多南方地区是水稻栽插的又一关键时期，有许多与之相关的农谚，如"立夏小满正栽秧""秧奔小满，谷奔秋"等。此时，降雨的变化、雨水的盈缺会直接影响到水稻的生长。"小满不满，干断田坎""小满不满，芒种不管"等农谚，都形象地表达了小满降雨的重要性。

Lesser Fullness of Grain is the key period for planting rice in south China including Sichuan region. There are lots of farmer's proverbs, like "the Beginning of Summer and Lesser Fullness of Grain are right time to plant rice", "plant rice in Lesser Fullness of Grain and harvest rice in autumn". In this period, the change and amount of rain fall will affect the growth of rice. There are farmer's proverbs, like "lack of rain fall in Lesser Fullness of Grain will cause the drought in future", and "lack of rain fall in Lesser Fullness of Grain will affect the production in Grain in Beard, which have described the importance of the rain fall in Lesser Fullness of Grain.

这一时节，四川地区的一些特色食材已开始成熟、收获。在资阳，此时的小龙虾已长得肥美诱人，进入大量捕捞季节，经川菜厨师之手而变成一道道色泽红亮、味型多样的龙虾菜肴，成为四川民众夏季宵夜不可或缺的美食。如今，资阳市雁江区中和镇已成为远近闻名的小龙虾之乡。此外，四川的车厘子在此时也大量上市，它分布广泛，品种多，产量高，体型圆润、色泽亮丽、甘甜多汁，既可直接食用，也可做成菜肴。

In this solar term, some special food ingredients in Sichuan region has grown up and can be harvested. In Ziyang City, it is the good time to catch crayfish. Sichuan chefs can make all kinds of bright red crayfish dishes with different tastes, which are the cores of Sichuan people's night snacks in summer. Nowadays, Zhonghe Town, Yanjiang District, Ziyang City has become the main producing area of crayfish. Furthermore, Sichuan big cherries are on the market too. The big cherry with high productivity is widespread. It is good looking, sweet and juicy, can be eaten fresh or cooked.

饮食养生 | DIETARY REGIMEN

小满过后，气温逐渐升高，雨水不断增多，天气越发闷热潮湿。这时，人们容易感染“湿邪”，主要表现为头身困重、舌苔白腻、食欲不振、嗜睡困倦等症状。与此同时，人们往往通过喝冷饮来消暑降温，而冷饮过量容易损害人体的阳气，也常引发老年人和孩子的肠胃不适、腹泻腹痛等病症。小满养生，要避免过量食用生冷食物，少食动物脂肪、海腥鱼类、葱、蒜、芥末、辣椒及各种发物，以免生湿助湿，而应以素为主，常吃有清热祛湿作用的食物，如赤小豆、薏米、冬瓜、丝瓜、黄瓜、黄花菜、水芹、草鱼、鲫鱼等，也可适量选用鸭肉、绿豆，以利清热养阴、解毒。资阳市盛产的小龙虾富含镁元素，有调节心脏活动、防止动脉硬化、安心养神之功效。

After Lesser Fullness of Grain, the temperature is getting high and the rain fall is increasing. The climate has become sultry and humid. People are easily infected by damp. People will feel tired, lack of appetite and sleepy. Normally, we can take cold drinks to relieve summer heat, but too much cold drinks will harm the yang qi in human body. It will cause gastrointestinal discomfort, abdomen pain with diarrhea. In Lesser Fullness of Grain, we should avoid eating too much cold food, animal fat, seafood, green onion, garlic, mustard and chilly. We should mainly eat vegetarian food, and food which can clear heat and damp, like red bean, pearl barley, winter melon, towel gourd, cucumber, day lily, cress, grass carp, crucian, etc. We can also eat some duck meat and mung bean, which can clear heat and relieve toxicity. Crayfish from Ziyang City are full of magnesium element, which can regulate cardiac system, prevent arteriosclerosis and calm the nerves.

小满美食 | FINE FOODS IN LESSER FULLNESS OF GRAIN

小满时节，旧粮已尽，新粮未收，正所谓青黄不接，而此时，苦菜疯长，恰好解决了粮食短缺的问题，由此，吃苦菜成为长久以来小满节气独特的食俗和传统。《周书》曰：“小满之日苦菜秀。”《诗经·邶风·谷风》言：“采苦采苦，首阳之下。”红军长征之时吃苦菜充饥，因此又被称为“红军菜”。据明朝李时珍《本草纲目》载：“苦菜，久服，安心益气，轻身耐老。”苦菜具有清热解毒、预防湿疮、轻身耐老的功效。如今，人们吃苦菜的方式较多，既可烫熟凉拌，也可挤出苦菜汁做面，还可清炒、做汤、煮粥，苦菜已成为小满时节恰逢其时的美食。

Lesser Fullness of Grain, the lately harvest grain has been eaten up and the newly harvest grain has not come yet. It is a period of temporary shortage. However, the fast-growing sowthistle can solve this problem. Eating sowthistle has become a special eat custom and tradition since long ago. According to *Book of Zhou*, sowthistle has become the star of Lesser Fullness of Grain day. *The Book of Songs · Beifeng · Gufeng* has mentioned about sowthistle as well. People call sowthistle Red Army Vegetable because that red army made it main course during the Long March. *Compendium of Materia Medica* says that keeping on eating sowthistle will clam the mind and prevent aging. Sowthistle has the function of clearing heat, removing toxicity and preventing eczema. Today, there are lots of way of eating sowthistle, like eating in sauce, frying, making soup, making noodles and making porridge. Sowthistle has become the right fine food in Lesser Fullness of Grain.

双味小龙虾

Double Flavor Crayfish

食材配方

香辣小龙虾： 小龙虾1000克　水发魔芋300克　洋葱条150克　黄瓜条150克　干辣椒节40克　花椒20克　郫县豆瓣20克　火锅底料40克　泡辣椒末30克　野山椒末40克　小米辣末60克　鸡精3克　味精3克　白糖5克　香料粉15克　醋10毫升　藤椒油50毫升　食用油2000毫升（约耗120毫升）

红糖糍粑小龙虾： 小龙虾500克　糍粑250克　红糖100克　白糖100克　酥花生碎15克　熟芝麻10克　清水200毫升

制作工艺

1.香辣小龙虾

① 小龙虾去头、虾脑、小脚和虾线，在背上片一刀，入160℃的油中炸至成熟后捞出。

② 锅置火上，入油烧至120℃时，放入干辣椒节、花椒、郫县豆瓣、泡辣椒末、野山椒末、小米辣末、火锅底料炒香，掺入清水烧沸后入小龙虾、水发魔芋、洋葱条、黄瓜条、白糖、鸡精、味精、香料粉、醋、藤椒油烧至入味，待收汁浓稠后起锅装盘即成。

2.红糖糍粑小龙虾

① 小龙虾去头、虾脑、小脚和虾线；糍粑切为约2厘米见方的丁；红糖切细。

② 锅置火上，入油烧至160℃时，放入小龙虾炸熟后捞出，再入糍粑炸至外表酥脆后捞出。

③ 锅置火上，注入清水、白糖、红糖，用小火熬制起小泡时放入小龙虾、糍粑丁炒匀，出锅装盘，撒上酥花生碎、熟芝麻即成。

风味特色

成菜色泽红亮，肉质细嫩、有弹性，麻辣与甜香并列，可食性强。

创意设计

资阳的雁江区中和镇盛产小龙虾，当地人最传统、最喜爱的菜肴就是香辣小龙虾，还为此研发出红糖糍粑小龙虾、鱼香小龙虾等十余个品种。此菜将香辣小龙虾与红糖糍粑小龙虾结合而成，麻辣与甘甜结合，可同时兼顾喜食辣与不食辣的人群需求。在养生方面，小龙虾补肾壮阳、滋阴息风，而无论是辣椒还是红糖都属温性，与小龙虾同用，可以温阳养阳，但阴虚火旺及皮肤病患者需慎食。

Ingredients

Spicy Crayfish: 1,000g crayfish; 300g water-soaked konjac jelly; 150g onions, shredded; 150g cucumbers, shredded; 40g dried chilies, segmented; 20g Sichuan peppercorns; 20g Pixian chili bean paste; 40g hotpot soup base; 30g pickled chilies, finely chopped; 40g mountain chilies, finely chopped; 60g birds' eye chilies, finely chopped; 3g chicken essence; 3g MSG; 5g sugar; 15g spice powder; 10ml vinegar; 50ml green Sichuan pepper oil; 2,000ml cooking oil (about 120ml to be consumed)

Brown Sugar Crayfish with Ciba Rice Cake: 500g crayfish; 250g Ciba rice cake; 100g brown sugar; 100g sugar; 15g fried peanuts, finely chopped; 10g roasted sesames; 200ml water

Preparation

Spicy Crayfish

1. Remove the head, brain, feet and vein of crayfish. Cut along their backs, and deep fry in 160℃ oil till cooked through.
2. Heat oil in a wok to 120℃, add dried chilies, Sichuan peppercorns, Pixian chili bean paste, pickled chilies, bird's eye chilies, mountain chilies and hotspot soup base to stir fry until aromatic. Pour in water, bring to a boil, and add the crayfish, konjac jelly, onions, cucumbers, sugar, chicken essence, MSG, spice powder, vinegar, and green Sichuan pepper oil to braise until the crayfish is flavored and the sauce is reduced. Transfer to a serving dish.

Brown Sugar Crayfish with Ciba Rice Cake

1. Remove the head, brain, feet and vein of crayfish. Cut the Ciba rice cake into 2cm dices. Finely chop the brown sugar.
2. Heat oil in a wok to 160℃, deep fry the crayfish till cooked through, and remove. Deep fry the rice cake till crispy.
3. Heat a wok over a low flame, add water, sugar, brown sugar till there are small bubbles in the sauce. Add the crayfish and rice cake, stir well, remove and transfer to a serving dish. Sprinkle with peanuts and sesames.

Features

alluring reddish color; spicy and sweet tastes; tender and chewy meat

Notes

Zhonghe Town of Yanjiang District of Ziyang City is famous for crayfish farming. Spicy Crayfish is local people's favorite. They also have developed over 10 crayfish dishes, such as Brown Sugar Crayfish with Ciba Rice Cake, Fish-flavor Crayfish and so on. Combining Spicy Crayfish with Brown Sugar Crayfish, this double flavor dish can meet the customer demands for both spicy and non-spicy foods. As for food therapy, crayfish has the nature of tonifying the kidney, strengthening yang and enriching yin. Chilies and brown sugar belong to warm nature foods, and can warm and nourish yang when combined with crayfish. However, this dish is not suitable for people who have skin diseases or suffer from inner heat.

冬瓜鸭方

Braised Duck with Winter Melon

食材配方

冬瓜500克　麻鸭肉300克　虫草花20克　红曲米100克　八角2个　姜片15克　葱段25克　食盐10克　料酒20毫升　白糖1克　鸡精2克　鸡汁4毫升　鲜汤1000毫升　芝麻油6毫升　水淀粉30克　食用油80毫升

制作工艺

① 冬瓜去皮后切成圆圈，在表面用U型戳刀戳成花形，入笼蒸熟后装盘备用。

② 鸭肉切成约2厘米见方的块，焯水后备用；红曲米入沸水中浸泡30分钟，取红曲米水备用。

③ 锅置火上，入油烧至160℃时，放入鸭肉、八角、姜片、葱段煸炒出香，再入鲜汤、红曲米水、食盐、料酒、白糖烧沸，之后改用小火烧至鸭肉软熟，捞出姜、葱、八角，放入鸡精、鸡汁、芝麻油、水淀粉，收汁至浓稠后舀入冬瓜里。

④ 锅置火上，入油烧至120℃时，放入姜片、葱段炒香，掺鲜汤烧沸出味后捞出姜、葱，入虫草花、鸡精、鸡汁、芝麻油、水淀粉收汁至浓稠，起锅浇淋在冬瓜上即成。

风味特色

色泽红白分明，冬瓜细嫩，鸭肉软熟，味道咸鲜香浓。

创意设计

麻鸭是四川特产的优质食材，而夏季食鸭是四川民间的饮食传统之一。此菜以麻鸭肉与当季出产的冬瓜为食材制作而成，与节令非常吻合。在养生方面，冬瓜味甘、性寒，有消热利水、消肿祛湿等作用，鸭肉清热养阴，两者配伍成菜，既能祛除小满时节人体容易感染的“湿邪”，又能避免伤阴耗气。此外，冬瓜还有助于促进新陈代谢、减肥、缓解糖尿病症。

Ingredients

500g winter melon; 300g duck; 20g cordyceps flowers; 100g red mold rice; 2g star aniseeds; 15g ginger, sliced; 25g spring onions, segmented; 10g salt; 20ml Shaoxing cooking wine; 1g sugar; 2g chicken essence; 4ml chicken bouillon; 1,000ml stock; 6ml sesame oil; 30g average batter; 80ml cooking oil

Preparation

1. Cut the winter melon into ring shape and peel. Use U-shape knife to cut flower patterns on the surface. Steam till cooked through, and transfer to a serving dish.
2. Chop the duck into 2cm cubes and blanch. Soak red mold rice in boiling water for 30 minutes, remove and save the soaking water.
3. Heat oil in a wok to 160℃, add the duck, star aniseeds, ginger, spring onions to stir fry till aromatic. Add the stock, red mold rice soaking water, salt, cooking wine and sugar, bring to a boil, and simmer over a low flame until the duck is soft and cooked through. Ladle out the ginger, spring onions and star aniseeds. Add chicken essence, chicken bouillon, sesame oil and batter, continue to simmer to reduce the sauce, and transfer into the winter melon.
4. Heat oil in a wok to 120℃, and add ginger and spring onions to stir fry till aromatic. Add stock, bring to a boil, and remove the ginger and spring onions. Add cordyceps flowers, chicken essence, chicken bouillon, sesame oil and batter to thicken the sauce, and drizzle over the winter melon.

Features

contrasting red and white colors; tender and mild winter melon; savory taste

Notes

The local duck chosen in this dish is one of the quality specialties in Sichuan. As for local people, it is a tradition to have duck in summer. This dish uses duck and fresh winter melons of the season as the main ingredients. In food therapy, winter melon can clear heat, disperse swelling and resolve dampness. Duck meat has functions of clearing heat and nourishing yin. Using both ingredients together can ward off dampness and illness, prevent harm to yin and waste of qi during the period of Lesser Fullness of Grain. Additionally, winter melon helps to improve metabolism, lose weight, and relieve the symptoms of diabetes.

车厘子冰粉
Ice Jelly with Cherries

食材配方

原味冰粉50克　开水5000毫升　新鲜车厘子200克　熟白芝麻20克　葡萄干50克　山楂碎50克　熟花生碎50克　蜂蜜250克　鲜薄荷叶50克

制作工艺

① 新鲜车厘子去核后榨成汁。

② 原味冰粉加入车厘子汁与开水搅匀，晾冷后入冰箱中冷藏，使其凝固成冰粉冻后取出。

③ 将冰粉冻舀入碗中，淋入蜂蜜，撒上熟白芝麻、葡萄干、熟花生碎和山楂碎，点缀上薄荷叶和车厘子即成。

风味特色

色泽丰富，质地细滑，甘甜香浓，冰凉爽口。

创意设计

冰粉是四川人夏季喜好的一款特色小吃。本品以车厘子与冰粉制成，更增添了车厘子的香味。在养生方面，冰粉原料含有大量膳食纤维，能增加饱腹感、降脂、降糖；车厘子能调中补气、益肾健脾，红色入心，正宜夏季食用，加上葡萄干、花生碎、山楂碎、芝麻等食材，不仅颜色鲜艳，而且富含不饱和脂肪酸等，也利于人体健康。

Ingredients

50g ice jelly powder; 5,000ml boiled water; 200g fresh cherries; 20g precooked white sesames; 50g raisins; 50g dried hawthorns, crushed; 50g precooked peanuts, crushed; 250g honey; 50g fresh mint leaves

Preparation

1. Core the fresh cherries, and process in a blender to make cherry juice.
2. Add cherry juice and boiling water to the ice jelly powder. Mix well, leave to cool, and refrigerate till firm.
3. Put the ice jelly in a bowl and pour some honey over the ice jelly, and sprinkle precooked white sesames, raisins, peanuts and dried hawthorns. Garnish with mint leaves and cherries.

Features

Beautiful colors; smooth, sweet and cold taste

Notes

Ice jelly is a special summer snack favored by people in Sichuan. This dish is made of ice jelly powder and cherries, which is full of strong cherry fragrance. In food therapy, the ice jelly powder is rich in dietary fiber, which gives people the feeling of fullness and reduces lipid and blood sugar. Cherry nourishes people's qi, tonifies the kidney and nourishes the spleen. It is suitable to eat cherries in summer. Meanwhile, raisins, peanuts, hawthorns and sesames are bright in color and rich in unsaturated fatty acid, which is good for people's health.

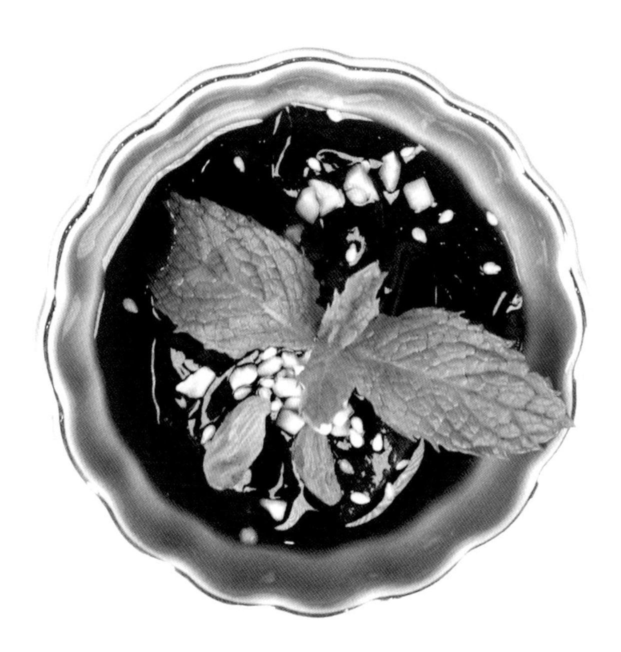

GRAIN IN BEARD
流水绿群芳摇落

芒种

溪流水绿 群芳摇落

咏廿四节气诗·芒种五月节

唐·元稹

芒种看今日，螗螂应节生。
彤云高下影，鴳鸟往来声。
渌沼莲花放，炎风暑雨情。
相逢问蚕麦，幸得称人情。

芒种物候 | PHENOLOGY IN GRAIN IN BEARD

芒种是二十四节气中的第九个节气。当太阳到达黄经75° 时为芒种，时间为每年的6月5日～7日。元朝吴澄《月令七十二候集解》言："五月节，谓有芒之种谷可稼种矣。"意思是说有芒作物，如大麦、小麦种子已经成熟，需要紧急抢收。芒种，也被称为"忙种"，意为赶紧去种，由此进入"三夏"大忙季节。

Grain in Beard is the ninth of the 24 Solar Terms. It begins when the sun reaches the celestial longitude of 75°. It usually begins from 5th to 7th June every year. In his book *A Collective Interpretation of the Seventy-Two Phenological Terms*, Wu Cheng (Yuan Dynasty 1271-1368) said that crops with awn, like barley and wheat, are ripe and need be harvest urgently in this period. Grain in Beard is also called busy planting (Mang Zhong), which means need plant the crops in a hurry. Since then, it is the three summer busy season.

我国古代将芒种后的十五天分为三候：一候螳螂生；二候鹏始鸣；三候反舌无声。芒种五日后，螳螂卵破壳而生出小螳螂；再过五日，喜阴的百劳鸟开始站立在枝头，感阴而啼鸣；再经五日，能够模仿其他鸟叫声的反舌鸟则不再鸣叫。

Ancient China has divided the 15 days after Grain in Beard into 3 pentads: in the first pentad, baby mantises go out of eggs; in the second pentad, shade-tolerant butcher bird start to sing on the branches; in the third pentad, mockingbird who can imitate other bird's songs stop singing.

食材生产 | PRODUCING FOOD INGREDIENTS

民间谚语曰："春争日，夏争时""栽秧割稻两头忙"。芒种时节是一年中农民最忙碌的季节，夏熟作物要及时收获，夏播秋收作物要马上播种，而春天播种的庄稼也要加强田间管理。在四川大部分地区，此时还常常需要将红薯进行移栽，若移栽过迟，不利于薯块膨大，会因此降低产量。

As the proverb goes, "in spring every day counts, in summer every hour counts" and "busy in harvesting early rice and planting summer rice simultaneourly". Grain in Beard is the busiest time for farmers in a year. They need harvest the summer maturing crops and plant autumn maturing corps urgently, and need manage the winter planting crops as well. In most of Sichuan region, sweet potato needs to be transplanted in this period, otherwise its output will be reduced.

这一时节，也是四川杨梅成熟、采摘的时节。四川杨梅形圆，色泽紫红，汁多肉嫩，食用方式多样，既可生食，也可做成杨梅干、酱、蜜饯等，还可酿酒。此外，四川西瓜在芒种时节也已成熟上市。西瓜在四川多地广泛种植，产量高、品种多，而资阳市迎接镇的西瓜更是远近闻名。

This solar item is the harvesting season for waxberry. Sichuan's waxberry is round-shaped, claret-colored, juicy and delicate. There are lots of ways of eating waxberry, like eating fresh, making sauce, making waxberry preserved and brewing wine. Morever, Sichuan's watermelon is on the market too. Watermelon here is widely planted. It has the characteristics of high output and more varieties. Watermelon from Yinjie Town, Ziyang City is known to all.

饮食养生 | DIETARY REGIMEN

芒种时节，暑热偏盛，降水较多，暑热与湿气交融混杂，呈现出潮湿闷热的节气特征。中医认为，人体与自然界相通应，湿易阻遏阳气，导致阳气郁滞、郁而为热，容易引起心烦燥热、精神不振、不思饮食、身体酸痛等不适之感。因此，芒种之时应注重清热祛湿、调养心神，饮食宜清淡，应少吃油腻味厚、辛辣上火的食物，减少肉类的食用量，可食用凉瓜、丝瓜、绿豆、冬瓜、西红柿、西瓜、桑椹、桃等时令蔬果，以及鲤鱼、鲫鱼、鸭肉等动物食材，既能补充营养，又可利湿清热、预防中暑。还可选用百合、莲子等药食两用之品以养心安神、固涩益气。此外，四川杨梅富含纤维素、矿质元素、维生素和一定量的蛋白质、脂肪、果胶及人体所需的八种氨基酸，具有生津止渴、和胃消食的功效。而资阳市出产的西瓜甘甜多汁，有美容养颜、消除倦怠、清热降火等功效，被誉为盛夏之王。

In Grain in Beard, it is a little bit hot and rainy. Humidity and stuffiness are this solar term's feature. TCM believes that human body is connected with the nature. Humidity will block yang qi and cause congestion and lassitude, body aches and loss of appetite. Therefore, people should clear summer heat, clear dampness and nurse the mind in Grain in Beard. We should take light food, avoid oily and spicy food, reduce too much meat. We can eat seasonal vegetables and fruits, like bitter ground melon, towel gourd, mung bean, winter melon, tomato, watermelon, mulberry, etc. Animal meat like carp, crucian and duck is suitable to eat. Taking those foods can supplement nutrition, clear dampness and heat, prevent heat stroke. We can also choose lily and lotus seed, which can be used as food and medicine, to nurse mind and tonify our energy. Furthermore, Sichuan waxberry is full of cellulose, minerals, vitamin, certain protein, fat, pectin and 8 amino acid, which can quench thirst, harmonize stomach and help digestion. Water melon from Ziyang City is sweet and juicy, which can maintain the beauty, relieve lassitude, clear heat, and is called the king of summer.

芒种美食 | FINE FOODS IN GRAIN IN BEARD

芒种时节最具代表性、最具悠久历史的食俗是煮梅。中国煮梅而食的习俗早在夏朝就已产生，《三国演义》中也记载有"青梅煮酒"的故事。每年芒种也是梅子成熟时节，但由于青梅味道过于酸涩，不宜直接食用，于是，人们便将梅子与白糖或食盐同煮后食用。有的地区还将梅子与甘草、山楂、冰糖一起煮水，制成酸梅汤饮用，以此清热解暑、健胃消食。

The most representative and historic eating custom in Grain in Beard is cooking plum. That tradition has been formed in Xia Dynasty (2070 BC-1600 BC). In *Romance of the Three Kingdoms*, there is the story of cooking green plum with liquor. Plum is ripe in Grain in Beard. However, green plum is too sour to eat directly, then people cook it with sugar or salt. In some places, people cook plum soup from plum, liquorice, haw, rock sugar and water. It can clear the summer heat, invigorate stomach and help digestion.

青梅话牛

Spiced Beef with Green Plum

食材配方

卤牛肉300克　青梅10颗　青梅汁10毫升　食盐1克　白糖60克　辣椒粉10克　花椒粉2克　酱油5毫升　芝麻油2毫升　清水50毫升　食用油1000毫升（约耗30毫升）

制作工艺

① 卤牛肉切成约1.2厘米见方的丁，入油锅中炸至表皮酥脆后捞出；青梅从中间切开，去掉果核。

② 锅置火上，入清水、白糖、青梅汁，用小火熬至浓稠后入辣椒粉、花椒粉、食盐、酱油、芝麻油炒匀，最后放入牛肉丁、青梅翻炒均匀，起锅装盘即成。

风味特色

成菜色泽棕红，牛肉质地酥脆，味道咸、甜、麻、辣、鲜、香、酸兼备。

创意设计

芒种时节，许多地方都有煮青梅而食的习俗，将梅子与糖或食盐同煮，风味独特。此菜选用青梅与牛肉搭配，两者兼为佐酒佳品，用怪味调制，不仅中和了青梅的涩味，还有较高的营养和食疗价值。其中，牛肉高蛋白、低脂肪，能强筋健骨、开胃祛湿；青梅生津利咽，并具有较强的清除自由基的能力和抗癌、抗疲劳、抗过敏等作用。

Ingredients

300g flavored beef (boiled in Sichuan-style spiced broth); 10 green plums; 10ml green plum juice; 1g salt; 60g sugar; 10g ground chilies; 2g ground Sichuan pepper; 5ml soy sauce; 2ml sesame oil; 50ml water; 1,000ml cooking oil (about 30ml to be consumed)

Preparation

1. Cut the flavored beef into 1.2cm cubes and deep fry till crispy on the outside. Halve the green plums and remover the cores.

2. Heat a wok over a low flame and add water, sugar and green plum juice. Stew till the soup is thickened. Add ground chilies, ground Sichuan pepper, salt, soy sauce and sesame oil. Stir fry and blend well. Add beef and green plums. Stir fry and blend well. Transfer to a serving dish.

Features

reddish brown color; crispy beef; savory, sweet, numbing, spicy, sour and aromatic flavors

Notes

At the time of Grain in Beard, there is a custom in many places that people cook green plums to eat. It has a special flavor to boil plums with sugar or salt in water together. This dish is made of green plums and beef, which are both suitable as an accompaniment for baijiu (an alcoholic drink). With strange flavor, the dish not only balances out the puckery taste of green plums, but also boasts high nutritional values and food therapy values. Among the ingredients, beef, rich in protein and low in fat, is able to strengthen the tendons and bones, stimulate appetite and clear damp. Green plum helps to promote body fluid and is able to scavenge free radicals. It also has anticancer, antifatigue and antiallergic effects.

银丝鳝鱼

Braised Eels with Pea Vermicelli

食材配方

鳝鱼200克　粉丝100克　香菜5克　蒜苗花3克　郫县豆瓣50克　姜米10克　蒜米10克　鸡精2克　味精2克　醋15毫升　白糖2克　胡椒粉1克　辣椒粉10克　鲜汤400毫升　食用油80毫升

制作工艺

①鳝鱼入锅，加盖煮沸2～3分钟后捞出，先用冷水漂凉，再用竹刀划成鳝丝；粉丝用沸水浸泡至透，捞出后用冷水浸泡备用。

②锅置火上，入油烧至120℃时，放入郫县豆瓣、辣椒粉、姜米、蒜米炒香，掺入鲜汤烧沸，入鳝鱼、粉丝、鸡精、味精、白糖、胡椒粉、醋烧沸，再用小火烧1～2分钟后起锅装盘，撒上香菜、蒜苗花即成。

风味特色

成菜色泽红亮，口感质地滑嫩，味道咸鲜酸辣。

创意设计

芒种，也称忙种，故有农谚说“栽秧割稻两头忙”。而在四川地区，此时正是捕捉水稻田中黄鳝的最佳时节。此菜以鳝鱼和安岳县特产的周礼粉丝为食材制成。在养生方面，鳝鱼能益气血、补肝肾、强筋骨、祛风湿，在芒种时节食用尤佳；粉丝调中益气，与鳝鱼搭配，更有助于祛湿健脾。鳝鱼除富含蛋白质外，其钙、磷、铁、维生素A及B族维生素含量也尤为突出，对降低血压、胆固醇，预防动脉硬化也有一定作用。

Ingredients

200g paddy eels; 100g Chinese pea vermicelli; 5g coriander; 3g baby leeks, finely chopped; 50g Pixian chili bean paste; 10g ginger, finely chopped; 10g garlic, finely chopped; 2g chicken essence; 2g MSG; 15ml vinegar; 2g sugar; 1g pepper; 10g ground chili; 400ml stock; 80ml cooking oil

Preparation

1. Blanch the paddy eels for 2-3 minutes, remove, cool in cold water, and cut into shreds. Soak the pea vermicelli in boiling water till it swells. Remove the pea vermicelli and soak in cold water.

2. Heat up the cooking oil to 120℃ in a wok, and add Pixian chili bean paste, ground chili, ginger and garlic. Stir fry till aromatic. Add the stock, bring to a boil, and add the eels, pea vermicelli, chicken essence, MSG, sugar, pepper and vinegar, and bring to a second boil. Braise over a low flame for 1-2 minutes. Transfer to a serving dish and sprinkle with coriander and baby leeks.

Features

bright and lustrous colors; smooth and tender eels and vermicelli; savory, sour and spicy tastes

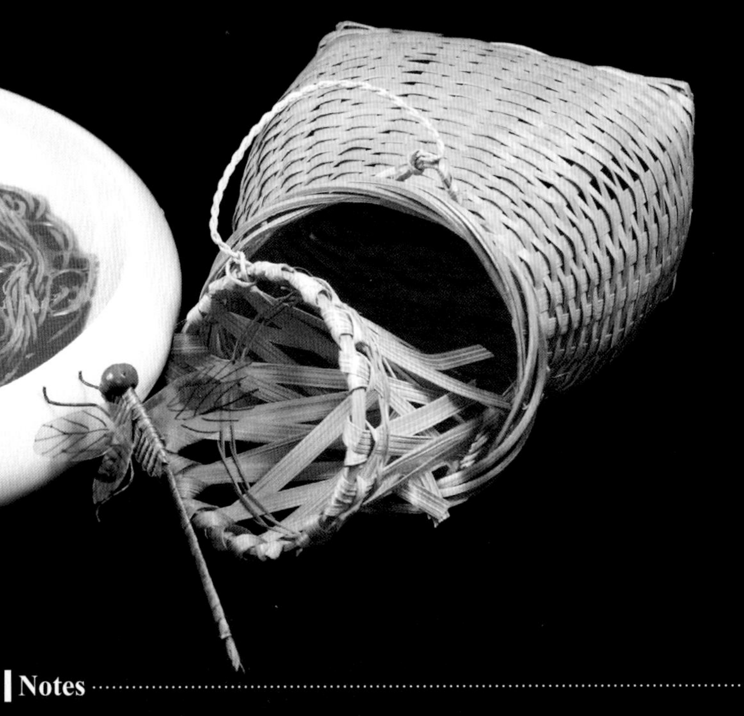

Notes

Grain in Beard is also called “a busy time for seeding”. In the farmer's proverb, it is very busy when famers transplant seedlings and harvest rice. At the time of Grain in Beard, people are ready to catch paddy eels in the rice field. This dish is made of paddy eels and Zhouli pea vermicelli, which is a specialty of Anyue County. In food therapy, paddy eel helps to nourish the blood, tonify the liver and the kidney, strengthen the tendons and bones and expel wind-damp. It is good to eat paddy eels around the Grain in Beard. Besides, pea vermicelli helps to nourish qi. Combining paddy eels with vermicelli, the dish is beneficial for tonifying the spleen and clearing dampness. In addition, paddy eel is rich in protein, calcium, phosphorus, iron, vitamin A and B-group vitamins, which help to lower blood pressure and cholesterol and prevent arteriosclerosis.

杨梅粽子

Zongzi in Waxberry Juice

食材配方

糯米500克　猪油50克　白糖100克　洗沙馅500克　新鲜杨梅果500克　鲜粽叶50张　专用粽子棉线1卷

制作工艺

① 将糯米入凉水中浸泡6小时后捞出，沥干水分，加入猪油和白糖拌匀；杨梅榨成汁。

② 鲜粽叶入沸水中煮5～6分钟至粽叶变柔软后捞出，入清水中漂冷。

③ 将粽叶卷成漏斗状，装入糯米和洗沙馅，再加盖一层糯米后拨平，包成粽子状，用棉线拴紧成生坯。

④ 锅置火上，放入粽子生坯和清水烧沸，然后改用中小火煮约2～3小时至糯米软糯时取出装盘，去掉粽叶，淋上杨梅汁即成。

风味特色

色泽红白鲜明，口感质地软糯，杨梅汁味道酸甜。

创意设计

芒种的时间有时与农历的端午节相邻，而粽子是端午节的传统美食。此款粽子是在传统豆沙粽子的基础上，搭配杨梅这一芒种时节的新鲜水果制作而成，令人回味无穷。在养生方面，杨梅清暑化痰、生津止渴，与补中益气的糯米一同做成粽子，不仅应季，更能益气生津。

Ingredients

500g glutinous rice; 50g lard; 100g sugar; 500g red bean paste; 500g fresh waxberries; 50 pieces of reed leaves; 1 reel of special cotton thread for Zongzi

Preparation

1. Soak the glutinous rice in cold water for 6 hours. Remove and drain. Blend lard and sugar and mix well. Squeeze the waxberries to make juice.
2. Blanch the reed leaves in boiling water for 5-6 minutes till soft. Remove and soak in cold water for use.
3. Shape the reed leaves into funnels, and add glutinous rice and red bean past. Add another layer of glutinous rice on top of the stuffing and wrap up to make Zongzi. Secure and fasten with cotton thread.
4. Heat a wok over the heat, add Zongzi and water, and bring to a boil. Turn to a medium-low flame. Boil for 2-3 hours till the glutinous rice turns to be sticky and soft. Transfer to a serving dish, remove the reed leaves and drizzle with the waxberry juice.

Features

white, bright and lustrous colors; soft and glutinous texture; sour and sweet tastes

Notes

The date of Grain in Beard is sometimes close to that of the Dragon Boat Festival, when the traditional food is Zongzi. This dish is made of traditional Zongzi (red bean paste stuffing) with fresh waxberry juice, which makes people appreciate every bite of it. In food therapy, waxberry helps to clear summer-heat, reduce phlegm, produce saliva and slake thirst. Glutinous rice helps to nourish qi. Combining these two ingredients, Zongzi is suitable for the Grain in Beard and good for tonifying qi and producing saliva.

THE SUMMER SOLSTICE

夏至

荷气初展 夏蝉始鸣

夏至过东市二绝（其一）

宋·洪咨夔

涨落平溪水见沙，绿阴两岸市人家。晚风来去吹香远，蕺蕺冬青几树花。

夏至物候 | PHENOLOGY IN THE SUMMER SOLSTICE

夏至是二十四节气中的第十个节气，也是被我国先祖最早确立的节气之一。当太阳到达黄经90° 时为夏至，时间为每年的6月21日~22日。《恪遵宪度抄本》记载："日北至，日长之至，日影短至，故曰夏至。"夏至，太阳几乎直射北回归线，是北半球一年中白昼最长、夜晚最短的一天。

The Summer Solstice is the tenth of the 24 Solar Terms, and one of the earliest solar terms confirmed by our ancestors. It begins when the sun reaches the celestial longitude of 90°. It usually begins from 21st to 22nd June every year. *Transcript of Ke Zun Xian Du* said, "in this period, day time is the longest and shadow is the shortest in the year; hence it is called the Summer Solstice ". In the Summer Solstice, sun is almost directly above the Tropic of Cancer. On the day of the Summer Solstice in the Northern Hemisphere, the day time is the longest and the night time is the shortest in a year.

我国古代将夏至后的十五天分为三候：一候鹿角解；二候蝉始鸣；三候半夏生。夏至五日后"鹿角解"，是因为古人认为鹿角朝前生，所以属阳。夏至日阴气生而阳气始衰，所以阳性的鹿角便慢慢开始脱落；再过五日，蛰伏已久的知了已感知到季节的更替，开始化蛹生翼、破土而出，雄性的知了在树间鼓翅而鸣，难怪"蝉"又被称为"知了"；再经五日，亲水喜阴的半夏开始生根萌芽，蓬勃生长。夏至日的到来，意味着夏天过半，古人故将这种颇具代表意义的植物称作"半夏"。由此可见，随着仲夏的来临，喜阴的植物开始滋长，而阳性植物则开始逐渐走向衰退。因此，夏至可以被视为自然界阴阳两气的分水岭。

Ancient China has divided the 15 days after the Summer Solstice into 3 pentads. In the first pentad, antler with the nature of yang begins to fall away because the yang qi is getting weak and ying qi is getting strong in the Summer Solstice. In the second pentad, dormant cicada is chirping on branches. In the third pentad, water-loving and shade-loving pinellia ternata starts to sprout and grow quickly. The arrival of the day of the Summer Solstice means it's halfway through summer, hence we call pinellia ternata "Half Summer" in Chinese. As mid-summer comes, shade-loving plants begin to grow and light demanding plants begin to weaken. Therefore, the Summer Solstice can be regarded as the watershed between yang qi and ying qi in the nature.

食材生产 | PRODUCING FOOD INGREDIENTS

夏至时节，温度高，日光足，雨水多，是四川地区各种农作物的快速生产期。此时，大豆、玉米等夏播秋收作物刚刚发芽，易遭受病虫侵蚀，再加之杂草易生，更需加强田间管理，及时施肥、打药。此时，四川多种蔬菜进入采摘期，如茄子、南瓜、苦瓜、丝瓜、土豆、四季豆、豇豆、嫩圆瓜、玉米等，为川菜制作提供了丰富的原材料。同时，桃子已经成熟上市。四川地区的桃子种植广、品种丰、产量大，龙泉、简阳等地的桃子尤其出名，个头圆滑、表皮红润、气味芳香、口感甜美。

In the Summer Solstice, the temperature is high, the sunshine is abundant and the rain fall is adequate. It is the rapid growth period of crops in Sichuan region. Crops like soybean, corn which sown in Summer just sprout, and easy to suffer from pest. Then people need to strengthen the filed management, to weed, and

provide fertilizer and spray insecticide. In this period, lots of vegetables in Sichuan are in picking period, like eggplant, pumpkin, bitter gourd, towel gourd, potato, kidney bean, cowpea, tender round melon, corn, etc, which can provide abundant materials to Sichuan cuisine. Meanwhile, peach is on the market. Peach in Sichuan region has the characteristics of wide plant, rich in varieties, and high output. Peaches from Longquan and Jianyang are especially famous, which are big and round, red colored, fragrant and sweet.

饮食养生 | DIETARY REGIMEN

夏至时节，人体的阳气最为充盛，心火当令，需注重保养心气、预防中暑。在饮食养生方面，以清补健脾、祛暑化湿为主要原则，宜食用清淡食物，忌油腻味厚之物，以免化热生风、引发疔疮。可以多食五谷杂粮以壮体，如小米性凉、玉米祛湿，还可适当吃一些苦味和酸味食物，以利清热滋阴，如绿豆、赤小豆、冬瓜、丝瓜、苦瓜等能清热祛湿；兔肉、鸭肉能凉血滋阴；苦瓜、苦苣、蒲公英、枸杞苗等能清解暑热；乌梅、山楂等酸味食物能开胃生津。代表菜品有南瓜薏米汤（或点）、三豆汤（红豆、白扁豆、绿豆）、陈皮冬瓜老鸭汤或炖品等。同时，不可贪食热性食物及肥甘厚味之物，还应少吃煎炸食品，也不可过多食寒凉食物，以免损伤脾胃。

In the Summer Solstice, yang qi in human body is the strongest in a year. Hence, people should pay attention to the heart energy and prevent sun stroke. Eating should follow the principles of tonifying spleen, dispelling humidity, clearing summer heat. In order to prevent stroke and furunculosis, it is better to take light foods, avoid oily and heavy foods. People can take more whole grains to strengthen the physique, like millet with the cold nature, corn with the function of dispelling humidity. We can eat some bitter and sour food to clear heat, like mung bean, red bean, winter melon, towel gourd and bitter gourd, which can clear summer heat and dispelling humidity. We can take rabbit meat and duck meet which can cool blood. We can take bitter gourd, endive, dandelion, wolfberry seedling, which can clear summer heat. We can eat dark plum and haw which can make us have good appetite. The representative dishes are Pumpkin and Pearl Barley Soup, Three Beans Soup (red bean, white lentil, mung bean), Dried Orange Peel Winter Melon and Old Duck Soup. Moreover, people should not eat hot, heavy and oily food, should eat less fried foods to avoid furunculosis, should take less cold food to protect spleen and stomach.

夏至美食 | FINE FOODS IN THE SUMMER SOLSTICE

夏至之后，太阳直射地面的位置逐渐南移，北半球逐渐昼短夜长。因此，民间广泛流传着“吃过夏至面，一天短一线”“冬至饺子，夏至面”之说，故在我国许多地区都有在夏至日吃面的习俗。《帝京岁时纪胜》载“京师于是日（夏至）家家俱食冷淘面”。夏至面在不同的地区有不同的吃法，如老北京人喜欢在夏至吃凉面和炸酱面，而在江南一带，一般会吃阳春面、过桥面及麻油凉拌面等。四川许多地区在夏至常常吃凉面，即将面条入沸水中煮熟，捞出沥干水分，淋上熟菜油拌匀，摊开晾凉，用焯水后的绿豆芽垫底，加入食盐、酱油、糖、醋、芝麻油、花椒油、辣椒油、蒜泥、葱花等调料制成，口感爽滑、麻辣鲜香，且具有健胃补脾、化阴生津、消暑解毒的功效。此外，在浙江绍兴一带，还有在夏至日吃“圆糊醮”的习俗。圆糊醮是将麦粉调成糊状，再佐以蔬菜，摊成圆饼后直接食用，也可以卷食。当地民间说“夏至吃了圆糊醮，踩得石头咕咕叫”，这也代表了当地人对强身健体的追求。过去，许多农民不仅把圆糊醮作为夏至美食，还将其用竹签穿好，插于稻田的缺口流水处，并焚香祭祀，以祈求时和年丰。

After the Summer Solstice, the position where the sun shines directly on the ground has moved towards south gradually. Day time has been shortened and night time has been extended in the Northern Hemisphere. As the saying goes, “after eating noodles in the Summer Solstice, the day time is being shortened everyday”, and “eat dumpling in the Winter Solstice, and eat noodles in the Summer Solstice”. In many parts of China, there is a custom of eating noodles on the day of the Summer Solstice. *Record of Festivals in Empire Capital* has said, “every family eats cold noodles in Capital on the day of the Summer Solstice.” There are different ways of eating that noodles in different places. Beijing people love cold noodle and meat sauce noodle. In south of Yangtze River, people normally eat plain noodle, noodle with shrimp, minced pork and egg, sesame oil noodle. Lots of parts in Sichuan prefer to cold noodles. People boil the noodle first, drain it, mix with rape oil, add spices (salt, soy sauce, sugar, vinegar, sesame oil, Sichuan pepper oil, chili oil, mashed garlic, chopped green onion), and then eat it with blanched mung bean seedlings. It tastes smooth, fresh and delicious, and can nourish stomach, tonify spleen, engender liquor, clear the summer heat and relieve toxicity. In Shaoxing, Zhejiang, there is a custom of eating pancake. People make flours into paste, mix it with vegetables and then make pancakes. As the local proverb says that eating this pancake can keep your body healthy. In the past, many farmers not only ate this type of pancake, but also put it in rice fields with burning incense sticks and candles to pray for harvest.

藿香乌鱼花

Huoxiang Snakehead

食材配方

乌鱼750克　鸡蛋清50毫升　淀粉50克　藿香碎100克　食盐4克　姜片5克　葱段10克　葱花50克　小米椒50克　仔姜50克　海鲜酱油20毫升　黄豆酱油10毫升　料酒10毫升　鸡精2克　味精2克　藤椒油10毫升　清汤750毫升　食用油1000毫升（约耗50毫升）

制作工艺

① 乌鱼取净肉，先剞成“十”字花刀，再切成约5厘米见方的块，加入食盐、姜片、葱段、料酒码味10分钟。

② 鸡蛋清与淀粉拌匀成蛋清淀粉；小米椒、仔姜切成颗粒。

③ 锅置火上，入油烧至150℃时，将乌鱼花均匀粘裹上蛋清淀粉入锅炸熟后捞出，摆放在盘中备用。

④ 锅置火上，掺入清汤，放入小米椒粒、仔姜粒、海鲜酱油、黄豆酱油、鸡精、味精、藤椒油烧沸制成味汁，起锅浇淋在乌鱼花上，最后撒上葱花、藿香碎即成。

风味特色

成菜色泽自然，肉质细嫩，具有弹性，味道咸鲜麻辣，藿香味突出。

创意设计

资阳所在的沱江流域大量出产乌鱼。此菜以乌鱼为主要食材制成，因调味时重用藿香而得名。养生方面，乌鱼能调中开胃，藿香可祛暑解表、化湿和胃，佐以仔姜、辣椒、花椒更能开胃祛湿，宜于常人夏至时节食用，但体质燥热者慎食。此外，藿香、仔姜、辣椒还有清除自由基、抗衰老、抗癌等作用。

Ingredients

750g snakehead fish; 50ml egg white; 50g cornstarch; 100g huoxiang herbs (agastache rugosus), finely chopped; 4g salt; 5g ginger, sliced; 10g spring onions, segmented; 50g spring onions, finely chopped; 50g bird' eye chilies; 50g tender ginger; 20ml seafood sauce; 10ml soy sauce; 10ml Shaoxing cooking wine; 2g chicken essence; 2g MSG; 10ml green Sichuan pepper oil; 750ml stock; 1,000ml cooking oil (about 50ml to be consumed)

Preparation

1. Get the flesh of snakehead and cut crosses on it. Cut into 5cm chunks. Add salt, ginger, spring onion segments and Shaoxing cooking wine, and marinate for 10 minutes.
2. Mix egg white and cornstarch to make a paste. Finely chop the bird's eye chilies and tender ginger.
3. Heat oil in a wok till 150℃. Roll the snakehead chunks in the paste, deep fry till cooked through, and ladle out to a plate.
4. Place a wok over the flame, add the stock, bird's eye chilies, tender ginger, seafood sauce, soy sauce, chicken essence, MSG and green Sichuan pepper oil, bring to a boil, and pour over the fish. Sprinkle with chopped spring onions and huoxiang herbs.

Features

tender and slightly springy fish; salty and spicy tastes; a special aroma from huoxiang herb

Notes

Snakehead fish abounds in the Tuojiang River in Ziyang. This dish is great for people's health. Snakehead whets the appetite while huoxiang herbs help to dispel summer heat, remove dampness, and restore normal functioning of the stomach. Especially in summer, tender ginger, chilies and Sichuan pepper in the dish enhance the function of removing dampness and developing appetite, but people suffering from inner heat should try the dish with caution. In addition, huoxiang herbs, tender ginger and chilies can scavenge free radicals, and fight aging and cancer.

鲜酿豆腐 | Stuffed Doufu

食材配方

嫩豆腐500克　猪肉碎100克　芹菜粒20克　折耳根粒20克　鲜汤100毫升　蚝油10毫升　鸡精2克　鸡饭老抽5毫升　东古一品鲜10毫升　姜葱水10毫升　食盐2克　鸡蛋清30毫升　干淀粉30克　水淀粉15克　食用油1000毫升（约耗80毫升）

制作工艺

①猪肉碎入盆，加入食盐、姜葱水、鸡蛋清、水淀粉搅匀成肉馅。

②嫩豆腐切成约4厘米见方的块，用勺子在其一面挖空，填入肉馅，用干淀粉封底。

③锅置火上，入油烧至180℃时，放入豆腐炸至表面色金黄后捞出装盘。

④锅置火上，入油烧至130℃时，放入猪肉碎炒香，入鲜汤、蚝油、东古一品鲜、鸡精、鸡饭老抽炒均，再入水淀粉收汁成二流芡，淋在豆腐上，最后撒上芹菜粒、折耳根粒即成。

风味特色

成菜色泽棕黄，豆腐细嫩，味道咸鲜，酱香浓郁。

创意设计

夏至时节，人体阳气最为充盛，宜食清淡、少油腻的食物。此菜在白嫩的豆腐中加入肉馅等制成，荤素搭配，高蛋白、低脂肪、多膳食纤维，又少油腻，且有嚼头。在养生方面，豆腐滋阴清热，猪肉滋阴润燥，芹菜清热利水，折耳根（鱼腥草）有清热解毒等作用。此菜清热效果明显，立夏时食用，能较好地祛除暑气。

Ingredients

500g doufu; 100g pork mince; 20g celery, finely chopped; 20g fish mint, finely chopped; 100ml stock; 10ml oyster sauce; 2g chicken essence; 5ml Chicken Rice dark soya sauce; 10ml Donggu soy sauce; 10ml ginger and spring onion juice; 2g salt; 30ml egg white; 30g cornstarch; 15g average batter; 1,000ml cooking oil (80ml to be consumed)

Preparation

1. Put pork mince in a basin, and add salt, ginger and spring onion juice, egg white and cornstarch. Blend well to make the stuffing.
2. Cut the doufu into 4cm cubes. Scoop a hole on one side and fill in the stuffing. Seal the hole with cornstarch.
3. Heat up the cooking oil to 180℃ in a wok. Add doufu and deep fry till golden brown on the surface. Remove and transfer to a serving dish.
4. Heat up the cooking oil to 130℃ in a wok. Add pork mince and stir fry till aromatic. Add the stock, oyster sauce, Donggu soy sauce, chicken essence, Chicken Rice dark soya sauce and average batter. Stir well till the sauce is thickened. Pour the sauce over doufu and sprinkle with finely chopped celery and fish mint.

Features

brownish yellow color; tender doufu; savory, delicate tastes with strong soy sauce flavor

Notes

At the time of the Summer Solstice, people's qi is the most adequate. Therefore, it is better to eat light and less greasy food. This dish is made of tender doufu stuffed with pork meat. It is a dish with balanced portion of vegetables and meat, rich in protein and dietary fiber, low in fat and less greasy, and chewy. In food therapy, doufu helps to nourish yin and clear inner heat; pork helps to nourish yin and moisten dryness; celery helps to clear away heat and promote body fluid; fish mint helps to clear away heat and detoxify the body. This dish is suitable for the time of the Summer Solstice, for it helps to dispel the summer heat.

五彩凉面

Colorful Cold Noodles

食材配方

湿细面条100克　鸡脯肉50克　绿豆芽15克　黄瓜15克　胡萝卜15克　紫甘蓝15克　复制酱油8毫升　食醋5毫升　酱油2毫升　食盐1克　味精0.5克　芝麻酱5克　蒜泥6克　葱花6克　芝麻油5毫升　红油辣椒25毫升　花椒油4毫升　熟菜油5毫升

制作工艺

① 黄瓜、紫甘蓝和胡萝卜分别切成细丝；绿豆芽入沸水焯熟后捞出晾冷；鸡脯肉煮熟后撕成细丝。

② 锅置火上，加水烧沸，入面条煮熟后捞出，沥干水分，加入熟菜油拌匀、抖散，晾冷后装入碗内，表面放上绿豆芽、黄瓜丝、胡萝卜丝、紫甘蓝丝和鸡脯肉丝。

③ 将复制酱油、食盐、酱油、味精、芝麻酱、花椒油、食醋和芝麻油搅匀制成调味汁，将其淋于凉面上，最后依次加入红油辣椒、蒜泥和葱花即成。

风味特色

菜丝色泽丰富，面条爽滑劲道，味道咸辣鲜香，回味甜酸。

创意设计

俗语说，“冬至馄饨，夏至面”，我国许多地区都有夏至吃面的习俗，而凉面是四川人夏季尤其喜好的小吃。本品在传统凉面的基础上，添加黄瓜、胡萝卜和紫甘蓝等蔬菜，既丰富了色泽，也增加了其营养价值。其中，黄瓜清热利水，绿豆芽清热消暑，胡萝卜补中益气，紫甘蓝开胃化痰，再辅之以鸡肉、面条，清暑之余更能补气健脾、爽口开胃。

Ingredients

100g wet thin noodles; 50g chicken breast; 15g mung bean sprouts; 15g cucumber; 15g carrot; 15g red cabbage; 8ml compound soy sauce; 5ml vinegar; 2ml soy sauce; 1g salt; 0.5g MSG; 5g sesame paste; 6g garlic, crushed; 6g spring onions, finely chopped; 5ml sesame oil; 25ml chili oil; 4ml Sichuan pepper oil; 5ml rapeseed oil

Preparation

1. Shred the cucumber, red cabbage and carrot into thin slivers. Blanch mung bean sprouts and leave to cool. Boil chicken breast till cooked through and hand shred into thin slivers.
2. Heat water in a wok until boiling. Boil the noodles until cooked through, remove and drain. Add rapeseed oil, mix well and shake till loose. Leave to cool, and transfer to a serving bowl. Cover the noodles with mung bean sprouts, cucumber, carrot, red cabbage and chicken breast.
3. Mix compound soy sauce, salt, soy sauce, MSG, sesame paste, Sichuan pepper oil, vinegar and sesame oil for seasoning. Drizzle the seasoning over the noodles, and add chili oil, garlic and spring onions in a successive order.

Features

colorful vegetable shreds; springy noodles; salty, spicy, savory, sweet and sour tastes

Notes

As a Chinese saying goes, “people prefer dumplings on the the Winter Solstice and noodles on the Summer Solstice”. People in many places in China has the custom of eating noodles on the Summer Soltice and cold noodles is one of the favorite cold dishes of Sichaun people. Based on traditional noodles, this dish uses more vegetables like cucumber, carrot and red cabbage, so it is more colorful and nutritious. Among all the vegetables, cucumber and mung bean sprouts clear body heat and the former also improves urination; carrot tonifies qi in human body; and red cabbage whets appetite. Coupled with chicken breast and noodles, this dish also replenishes qi, strengthens the spleen, clears body heat and whets appetite.

LESSER HEAT

小暑
晴日暖风 绿阴幽草
小暑六月节
唐·元稹
倏忽温风至，因循小暑来。
竹喧先觉雨，山暗已闻雷。
户牖深青霭，阶庭长绿苔。
鹰鹯新习学，蟋蟀莫相催。

小暑物候 | PHENOLOGY IN LESSER HEAT

小暑是二十四节气中的第十一个节气。当太阳到达黄经105° 时为小暑，时间为每年的7月6日~8日。元朝吴澄《月令七十二候集解》记载：“小暑，六月节”，“就热之中分为大小，月初为小，月中为大，今则热气犹小也。”暑，即酷热之意，小暑就是小热，此时还没有到达一年中最热的时候。

Lesser Heat is the eleventh of the 24 Solar Terms. It begins when the sun reaches the celestial longitude of 105°. It usually begins from 6th to 8th July every year. In his book *A Collective Interpretation of the Seventy-Two Phenological Terms*, Wu Cheng said that Lesser Heat is in June (lunar calendar). There is slight heat and great heat in this month. It is moderate at the beginning of month and great in the middle of the month. This solar term means moderate heat and has not come to the hottest time in a year.

我国古代将小暑后的十五天分为三候：一候温风至；二候蟋蟀居宇；三候鹰始鸷。小暑五日后，地上吹的风已开始炽热；再过五日，蟋蟀因炎热离开田野、到庭院阴角处避暑；再经五日，老鹰由于地面温度太高而开始飞到凉爽的高空中去捕猎。

Ancient China has divided the 15 days after Lesser Heat into 3 pentads: in the first pentad, wind begins to be hot; in the second pentad, cricket has left the field to court corner because of torridity; in the third pentad, eagles have flown to cool high altitude for hunting.

食材生产 | PRODUCING FOOD INGREDIENTS

小暑时节，四川地区进入雷电、暴雨较多的时期，有时伴着大风、冰雹等自然灾害的发生，在食材生产过程中，应特别注意预防自然灾害，不断加强田间管理，及时给生长中的水稻浇水、施肥，同时对玉米、大豆、红薯等遭受虫害的作物开展防治病虫工作。这一时期，仔姜开始大量上市。四川仔姜脆嫩少辣、汁多爽口，品质优良，尤其是资阳的鼎新仔姜远近闻名。

In Lesser Heat, there are more thunders and rainstorm with high wind and hail sometimes in Sichuan, which will cause natural catastrophes. During the period of producing food ingredients, people should pay special attention to natural disasters, strengthen fields management. One the one hand, people need irrigate and fertilize the growing rice; on the other hand, people should control pest for corn, soy bean and sweet potato. Tender ginger is on the market in this period. Sichuan tender ginger is crisp, less spicy, juicy, with high quality. Dingxin ginger in Ziyang is very popular.

饮食养生 | DIETARY REGIMEN

小暑时节，天气温度高，自然界中阳气旺盛，人体易出汗，俗话说“气随汗脱”，大量出汗易导致人体阳气受损，加之人们因食用大量寒凉食物，也会耗伤阳气，有悖于“春夏养阳”之道，因此要注意适度降温防暑，不宜出汗过多，不宜冷饮过度。同时，小暑的暑湿之邪易损伤脾胃，出现精神不振、浑身乏力、头晕目眩、食欲下降等症状。饮食上需清热利湿、健脾利湿、养心安神。除了可食用薏米、红豆、淡水鱼及各种瓜类祛暑除湿外，还可多选择黄色食物，如玉米、黄豆、柑橘、香蕉、枇杷、柠檬等，以入脾益气，也可适量食用百合滋阴润肺、养心安神，还可选用苦瓜、苦荞等。此外，民间俗语说“冬吃萝卜夏吃姜，不找医生开药方”，仔姜便是小暑时节养生的重要节令食材。用芡实、白扁豆、山药、橘皮、小米等做粥、做汤或面点小吃，用冬瓜、百合做菜和煮汤等，也是很好的选择。

In Lesser Heat, the temperature is high, and the yang qi in the nature is strong. In such weather, human body sweat easily. Chinese people believe that the yang qi will leak with sweat, and eating too much cold food will waste yang qi, which break the principle of nourishing energy in spring and summer. Therefore, we should prevent the summer heat, control the sweat, take less cold drink. Moreover, the summer humidity in Lesser Heat will damage spleen and stomach, and will cause lassitude, malaise, dizziness, lack of appetite. People should take foods that can clear summer heat, dispel humidity, tonify spleen and calm the mind. In addition to pearl barley, red bean, freshwater fish and all kinds of melons, people can eat more yellow foods, like corn, soy bean, orange, banana, loquat and lemon, which can tonify spleen and our energy. People can also eat proper lily, bitter gourd and buckwheat to tonify liver. As the proverb goes, “you do not need doctor if eating radish in winter and ginger in summer”. Tender ginger is the important seasonal food ingredients in Lesser Heat. It is a good option to make porridge, soup and pastry snacks with gorgon fruit, white lentil, Chinese yam, orange peel, millet. Making dishes and soups with winter melon and lily are good as well.

小暑美食 | FINE FOODS IN LESSER HEAT

自小暑开始，许多地方都有“尝新”的习俗，即品尝新收获的麦稻食品。农民将夏熟的稻谷碾成米，煮成米饭或酿制成新酒，先用来祭祀五谷大神和祖先，再与全家共同食用，借以表达人们感恩自然的馈赠和庆祝丰收的喜悦心情。有的地区则将新收获的小麦炒熟后磨成粉，加入白糖和水食用。

Since Lesser Heat, there has been a custom of tasting a fresh delicacy in many places. That means tasting food made of newly picked wheat and rice. Farmer will offer rice harvested in summer or liquor brew with that rice to God of Grains and ancestors firstly, then share with the whole family to thank nature and celebrate the harvest. People in some regions stir-fry the newly picked wheat and grind into flour, and eat with sugar and water.

小暑过后就进入了伏天，人们喜欢“吃伏面”来清凉消暑。南方地区吃伏面的习俗由来已久，南朝（梁）宗懔《荆楚岁时记》言：“六月伏日食汤饼，名为辟恶。”这里的汤饼就是热汤面。今天，人们吃伏面的方式变得多种多样，不仅食用汤面，还喜食干拌面、凉面等。

After Lesser Heat, there are dog days. People like to eat noodles in dog days to clear summer heat. It is a long time custom in the south. Zong Lin in the Southern Dynasty (502-557) has said in *Records of Festivals in Jingchu Region*, “people eat Tang Bing in June dog days to avoid evil.” Tang Bing is hot noodles with soup. Nowadays, there are lots of ways to eat dog days noodles. People take not only hot noodles with soup, but also dry noodles with sauce, cold noodles, etc.

双味仔姜

Double Flavor Tender Ginger

食材配方

甜酸味： 鲜仔姜200克　食盐5克　白醋5毫升　白糖30克　柠檬汁20毫升　味精2克

豆瓣味： 鲜仔姜200克　野山椒50克　食盐6克　泡菜水500毫升　临江寺豆瓣15克

制作工艺

① 甜酸味仔姜

仔姜切成梳子背块，加入食盐、白醋、白糖、柠檬汁、味精拌匀，入冰箱冷藏两小时。

② 豆瓣味仔姜

仔姜切成梳子背块，放入盆中，入野山椒、食盐、泡菜水浸泡两小时，捞出后装盘，放上临江寺豆瓣，与甜酸味仔姜一同装盘即成。

风味特色

色泽黄红相间，口感质地脆嫩，味道甜酸辣香。

创意设计

俗语说“冬吃萝卜夏吃姜，不找医生开药方”。先贤孔子在进餐中即“不撤姜食”，表现出其对姜的喜爱及食疗价值的肯定。而安岳鼎新出产的仔姜芽长筋少、质地脆嫩、辛辣中带有清香，是暑日食姜的最佳选择之一。此菜是在资阳民间流行的凉拌仔姜基础上改进而成。在养生方面，仔姜可解表散寒、温中止渴、化痰止咳，且有抑制细菌、保肝利胆等作用。

Ingredients

Sweet-Sour Flavor: 200g fresh tender ginger; 5g salt; 5ml white vinegar; 30g sugar; 20ml lemon juice; 2g MSG

Chili Bean Flavor: 200g fresh tender ginger; 50g mountain chilies; 6g salt; 500ml pickle brine; 15g Linjiangsi chili bean paste

Preparation

1. Sweet-Sour Flavor

Cut the tender ginger in the shape of comb back. Add the salt, white vinegar, sugar, lemon juice and MSG, mix well and refrigerate for two hours.

2. Chili Bean Flavor

Cut the tender ginger in the shape of comb back and transfer to a basin. Add the mountain chilies, salt and brine, and soak for two hours. Transfer to a serving dish, and top with Linjiangsi chili bean paste. Transfer the sweet-sour flavor tender ginger to the serving dish, and serve.

Features

light yellow and red in color; crispy and tender texture; mixed tastes of sweetness, sourness and spicincss

Notes

As a Chinese saying goes, people who eat radish in winter and ginger in summer don't see a doctor. Confucius, one of the great sages in ancient China, had ginger for every meal, which shows his favor for ginger and acknowledgement of its health care value. Tender ginger produced in Dingxin Township, Anyue County has less fibers and tastes cripsy, tender, fresh and spicy. Therefore, Dingxin ginger is one of the best ginger choices for summer. Double Flavor Tender Ginger is the improved version of Tender Ginger Salad, a popular folk dish in Ziyang. This dish is also great for people's health. Tender ginger relives the pores, clears cold in body, slakes thirst, prevents phlegm and coughing, controls bacteria and tonifies the spleen and the kidneys.

荷叶旱蒸滑肉

Steamed Pork in Lotus Leaves

食材配方

带皮五花肉200克　郫县豆瓣25克　料酒5毫升　姜米5克　味精1克
刀口花椒5克　红薯淀粉100克　荷叶1张

制作工艺

① 带皮五花肉切成片，加入郫县豆瓣、料酒、姜米、味精、刀口花椒拌匀，放入盘中，在每片肉上分别撒上红薯淀粉，用沸水浇淋定型。

② 荷叶放入小笼垫底，放入定型的肉片，再盖上荷叶，用旺火蒸熟即成。

成菜特色

滑肉色泽透亮，口感劲道，味道咸鲜麻辣，略带荷叶清香。

创意设计

荷叶是夏日的标志，碧绿晶莹，亭亭玉立，在小暑、大暑时节撑起一把把绿伞，给世间送来一片清凉与清香。荷叶旱蒸滑肉是在资阳传统名菜旱蒸滑肉的基础上用荷叶包裹后制成，既保留了滑肉透亮劲道的特点，又突出了荷叶的清香。在养生方面，猪肉滋阴润燥，搭配荷叶消暑祛湿，蒸制成菜更减油腻。此外，荷叶含有的生物碱、黄酮类及挥发油等成分，有抑菌、抗氧化、降脂、降胆固醇等作用。

Ingredients

200g pork belly; 25g Pixian chili bean paste; 5ml Shaoxing cooking wine; 5g ginger, finely chopped; 1g MSG; 5g Sichuan peppercorns, roasted and finely chopped; 100g sweet potato starch; 1 lotus leaf

Preparation

1. Slice the pork belly and add Pixian chili bean paste, Shaoxing cooking wine, ginger, MSG, chopped Sichuan peppercorns, mix well and transfer to a plate. Sprinkle sweet potato starch on every slice of pork, and pour over boiling water to turn the pork stiff.
2. Lay the lotus leaf on the bottom of a small steaming rack. Put the pork on the leave and cover with the extra lotus leaf. Steam over high heat till cooked through.

Features

crystal and springy slices of pork belly; salty and spicy taste; a fresh aroma of lotus leave.

Notes

Lotus leaves are one of the symbols of summer. They are verdant and graceful. In the height of summer, lotus leaves bring people refreshing cool. Steamed Pork in Lotus Leaves is an innovated version of the traditional dish Steamed Pork, using lotus leaves to wrap up the pork. The new cooking method keeps not only the elasticity of pork belly but also the fresh aroma of lotus leaves. This dish is also great for people's health. Pork nourishes yin and moistens body dryness, and lotus leaves remove body heat, clear dampness and cleanses the palate. As they have alkaloid, flavonoid, volatile oil and other elements, lotus leaves can fight bacteria and oxidation, and reduce lipid and cholesterol.

周礼伤心凉粉

Zhouli Pea Jelly

食材配方

豌豆300克　清水1000毫升　红油辣椒150毫升　花椒粉15克　姜汁20毫升　蒜泥20克　葱花50克　熟芝麻10克　酥花生碎30克　食盐2克　水淀粉30克

制作工艺

① 豌豆用冷水浸泡10个小时，磨成汁后过滤沉淀，将沉淀的精粉与部分水汁取出后入锅，另加清水搅拌至糊状，盛入容器中冷却后即成凉粉。

② 将食盐、水淀粉入锅调成咸酱汁。

③ 将凉粉切为长粗丝装入碗中，依次加入咸酱汁、红油辣椒、花椒粉、姜汁、蒜泥、熟芝麻、酥花生碎、葱花拌匀即成。

风味特色

成菜色泽亮白，质地爽滑软脆，味道咸鲜香辣。

创意设计

周礼伤心凉粉是资阳著名小吃品种，最宜暑日食用，既清凉开胃，又有食疗价值。其中，豌豆淀粉益气和中，姜汁、蒜泥、花椒粉皆能温中散寒，化湿下气，温阳护阳，避免凉粉性凉伤脾。此外，姜、蒜、花椒中的挥发油有抗菌、抗癌、抗氧化、抗衰老等作用。

Ingredients

300g peas; 1,000ml water; 150ml chili oil; 15g ground Sichuan pepper; 20ml ginger juice; 20g garlic, crushed; 50g spring onions, finely chopped; 10g roasted sesames; 30g crispy peanuts, crushed; 2g salt; 30g average batter

Preparation

1. Soak the peas in cold water for 10 hours, grind for juice and leave to settle. Transfer the sediments to a pot, add water, and stir till a paste has formed. Transfer the paste to a container, and leave to cool so that jelly forms.
2. Combine salt and average batter in a wok, and mix well to make the salty sauce.
3. Slice the jelly into long, thick strips and remove to a serving bowl. Add the salty sauce, chili oil, ground Sichuan pepper, ginger juice, garlic, sesames, peanuts and spring onions, mix well and serve.

Features

crystal white color; smooth, soft and crispy texture; salty and spicy tastes.

Notes

Zhouli Pea Jelly is a famous cold snack in Ziyang. It is good to have the dish in summer as it clears heat and whets appetite and has great health care value. Particularly, peas starch repenishes qi, and the volatile oil in ginger, garlic and Sichuan pepper prevents bacteria, cancer, oxidation and aging.

GREATER HEAT

大暑

清風不至火伞高张

大暑

宋·曾几

赤日几时过，
清风无处寻。
经书聊枕籍，
瓜李漫浮沉。
兰若静复静，
茅茨深又深。
炎蒸乃如许，
那更惜分阴。

大暑物候 | PHENOLOGY IN GREATER HEAT

大暑是二十四节气中的第十二个节气。当太阳到达黄经120° 时为大暑，时间为每年的7月22日~24日。元朝吴澄《月令七十二候集解》言：“大暑，六月中。暑，热也，就热之中分为大小，月初为小，月中为大，今则热气犹大也。”大暑正值三伏天的中伏，气温极高，酷暑难耐，民间谚语有“冷在三九，热在中伏”之说，正如俗话所言：“小暑大暑，有米不愿回家煮。”

Greater Heat is the twelfth of the 24 Solar Terms. It begins when the sun reaches the celestial longitude of 120°. It usually begins from 22nd to 24th July every year. In his book *A Collective Interpretation of the Seventy-Two Phenological Terms*, Wu Cheng said that Greater Heat is in the mid-June (lunar calendar). Summer is hot. It is slight hot at the beginning of this month and great hot in the middle of this month. Greater Heat is just in the middle of dog days. It is extremely hot then. As the proverbs go, “it is coldest in the third nine-day period after the Winter Solstice and it is hottest in middle of dog days” and “it is too hot in Lesser Heat and Greater Heat to cook at home”.

我国古代将大暑后的十五天分为三候：一候腐草为萤；二候土润溽暑；三候大雨时行。大暑五日后，腐草中的萤火虫卵化而出；再过五日，天气愈发闷热，土地更加潮湿；再经五日，大雨经常出现，暑热逐渐减弱。

Ancient China has divided the 15 days after Greater Heat into 3 pentads: in the first pentad, firefly comes out from the egg in rotten grass; in the second pentad, the weather is stuffier and the soil is more humid; in the third pentad, it is raining heavily sometimes and the summer heat begins to weaken.

食材生产 | PRODUCING FOOD INGREDIENTS

农谚说：“早稻抢日，晚稻抢时”“大暑不割禾，一天少一箩”。在大暑时节，四川地区进入抢种与抢收双季稻最忙碌、最艰苦的“双抢”时期，人们常常是天晴割稻、下雨插秧，以便在7月底前完成双晚稻插秧。这一时期，四川大部分地区荷花争相吐艳，荷叶接天碧绿。资阳市丹山镇每年都在此时节会举办盛大的荷花节，因为远近闻名，所以吸引了大量的游客到此赏荷、收荷、吃荷。荷全身是宝，荷花、荷叶、莲子、莲房、莲藕均可食用，其中，莲藕既可当作菜肴，又可制成藕粉食用，资阳的“天池藕粉”即是其中的名品。

As farmer’s proverb goes, farmers should seize time to plant and harvest rice. Greater Heat is the busiest period for Sichuan farmers to plant and harvest rice. People usually harvest the rice in fine days and transplant seedling in rainy days. In this period, most lotuses in Sichuan blossom. Danshan Town in Ziyang City will hold great Lotus Festival in this period every year. Lots of tourists have been attracted to appreciate lotus, harvest lotus and eat lotus there. Lotus petal, lotus leaf, lotus seed, lotus seed pot and lotus root are edible. Lotus root can be cooked directly and made into lotus root starch. Tianchi lotus root starch in Ziyang City is among the most famous.

饮食养生 | DIETARY REGIMEN

大暑正值六月中，此时为长夏之季和夏季之末，土壤中的濡润之水被暑热蒸腾为湿气，正所谓“暑必夹湿”。此时，湿邪与暑邪最易胶结而侵犯脾脏，故应在健脾护胃的同时注意祛湿。藿香芳香化湿，薄荷疏风清热、清利头目，荷叶消暑祛湿，均可适量食用，可将它们分别与其他食材一起烹制成菜点，如藿香鲫鱼、薄荷凉糕、荷叶粉蒸肉、荷叶肚包鸡等。同时，暑热易耗液伤津、损伤正气，导致食欲不振、精神萎靡，普通人的饮食宜选用性偏凉、微寒，并且味甘淡、酸或微苦的食物，如鸭肉、瘦猪肉、山药、薏苡仁、鲜藕、西红柿、冬瓜、柠檬等，而身体虚寒、脾阳虚弱的人宜食用鳝鱼、红枣、荔枝、椰肉等温补之品。

Greater Heat is in mid-June. It is the season of long summer and the end of summer. The water inside the soil has been evaporated by summer heat, hence the summer must be humid. At this moment, humid and summer heat will work together to damage spleen. People should pay attention to tonify spleen and dispelling humidity. Agastache rugosus can dispel humidity; mint can clear heat and mind; lotus leaf can clear summer heat and dispel humidity. People can make dishes with them and other food ingredients, like Agastache Rugosus Crucian, Mint Cool Cake, Lotus Steamed Pork with Rice Flour, Lotus Leaf with Pig Stomach and Chicken. Summer heat will consume body fluid, damage our energy, affect appetite, cause drowsy. People should choose sweet, sour and bitter foods with cold nature, like duck, lean pork, Chinese yam, coix seed, fresh lotus root, tomato, winter melon and lemon. People with cold nature and weak spleen need eat eel, red date, litchi and coconut meat, which are warm and nutritious.

大暑美食 | FINE FOODS IN GREATER HEAT

大暑时节的民间饮食习俗各地不一，南北不同，在我国南方一些地区有吃仙草、喝莲子汤的习俗，民间流传着“六月大暑吃仙草，活如神仙不会老”之说。所谓仙草，又名凉粉草、仙人草，有消热祛暑的功效，如今，四川的烧仙草冷饮店比比皆是，它是将凉粉草晒干后做成胶状，再浇入奶茶中饮用，有点儿类似于“龟苓膏”，冷热皆宜。莲子汤种类多样，主要有百合莲子汤、木瓜莲子汤、银耳莲子汤、猪心莲子汤等。莲子性味甘平，能补脾止泻、养心安神、益肾涩精。此外，四川一些地区还有吃冰汤圆以避暑消夏的习惯。每到盛夏，四川的一些大街小巷就会有冰汤圆的叫卖声。冰汤圆主要由汤圆、醪糟、红糖、冰沙等制成，味道酸甜、清爽宜人，具有健脾开胃、清热补虚、调血理气的功效。

Eating customs in Greater Heat are different between south and north. In south China, people has the custom to eat grass jelly and drink lotus seed soup. As the farmer’s proverb goes, “you can live as an immortal after eating grass jelly”. Grass jelly is called Chinese mesona or fairy grass, which can clear summer heat. Nowadays, there are so many cold drink shops selling grass jelly in Sichuan. People add grass jelly in cold or hot milky tea, just like herbal jelly. Lotus seed soup has various types, like Lily Lotus Seed Soup, Papaya Lotus Seed Soup, Tremella Lotus Seed Soup, Pig Heart Lotus Seed Soup, etc. Lotus seed taste sweet and neutral, can tonify spleen and stop diarrhea, calm mind, benefit kidney and arrest seminal emission. Moreover, there is a custom of eating iced Tangyuan in some parts of Sichuan. In every mid-summer, vendors sell iced Tangyuan in big streets and small alleys. Iced Tangyuan is made of Tangyuan, fermented glutinous rice, brown sugar, smoothie. It tastes sour and sweet, can make people feel refreshing and invigorating, can tonify spleen and stomach, clear summer heat and reconcile blood and energy.

折耳根白乌鱼汤

White Snakehead Soup with Fish Mint

食材配方

白乌鱼1尾（约600克）　土鸡500克　折耳根100克　姜片10克　葱段20克　料酒20毫升　胡椒粉1克　食盐10克　味精2克

制作工艺

①将白乌鱼、土鸡均斩成约4厘米见方的块，入沸水锅中焯水后备用；折耳根切成长约5厘米的段。

②锅置火上，加入清水和土鸡，下姜片、葱段、料酒、胡椒粉，先用大火烧沸，再改用小火炖两小时至鸡肉软熟，之后放入白乌鱼块、折耳根炖10分钟，最后下食盐、味精调好味即成。

成菜特色

成菜汤色清亮，鸡肉熟软，鱼肉细嫩，咸鲜中伴有折耳根香味。

创意设计

乐至白乌鱼是资阳市乐至县的特产，为国家地理标志产品。白乌鱼色白、无刺，肉质细嫩、鲜美，既营养又滋补，故有“鱼中珍品”之称。此菜将白乌鱼与土鸡、折耳根同炖，食疗价值更佳。其中，白乌鱼、土鸡皆为高蛋白、低脂肪的优质食材，益气开胃、滋补作用显著，与折耳根搭配，可补中有清，避免滋腻。此外，折耳根所含的挥发油成分，有增强免疫力及抗菌、抗病毒等作用。

Ingredients

1 white snakehead (about 600g); 500g free-range chicken; 100g fish mint; 10g ginger, sliced; 20g spring onions, segmented; 20ml Shaoxing cooking wine; 1g pepper; 10g salt; 2g MSG

Preparation

1. Cut the fish and chicken into 4cm cubes. Blanch the chicken in boiling water. Cut the fish mint into 5cm sections.
2. Heat a wok, and add water, chicken, ginger, spring onions, cooking wine and pepper. Bring to a boil, turn down the heat and simmer for two hours till the chicken is soft and cooked through. Add the fish and fish mint, simmer for another ten minutes, season with salt and MSG, and remove from the heat.

Features

clear soup and tender fish; savory tastes with a special fragrance from the fish mint

Innovation

Lezhi white snakehead, a local specialty of Lezhi County and a national geographic indication product, is known as delicacy of all fishes for it has few bones and its meat is tender, delicate and nutritious. Its food therapy benefits are doubled with free-range chicken and fish mint added. White snakehead and free-range chicken, which are high in protein but low in fat, increase qi circulation, boost the appetite, and invigorate the body. Fish mint, which has a cooling function, helps to reduce greasiness of the dish. Besides, the volatile oil contents in fish mint are antibiotic, antiviral and immunity-enhancing.

软炸荷花

Soft Fried Lutus Flowers

食材配方

鲜荷花瓣100克　脆浆糊100克　干辣椒粉碟1个　蜂蜜20克　酸梅酱10克　莲蓬1颗
食用油1000毫升（约耗30毫升）

制作工艺

① 鲜荷花瓣入淡盐水中洗净后凉干；蜂蜜、酸梅酱入碗兑成酱汁后装入小蝶。

② 锅置火上，入油烧至150℃时，将鲜荷花瓣挂上脆浆糊后入锅炸至金黄时捞出，摆放在有荷花、莲蓬造型的盘中，配上干辣椒粉碟、蜂蜜酸梅酱碟即成。

成菜特色

色泽金黄，造型美观，质地酥软，味道咸辣甜酸，荷花香味浓郁。

创意设计

荷花是夏日的花语。大暑时节，酷暑难耐，食欲不振，清雅、粉红的荷花会带来一份好心情。此时，资阳丹山的荷花节常常推出清香可口荷花菜品，让人流连忘返。此菜以荷花瓣为食材制成，造型为绽放的荷花，既凸显了其出污泥而不染的品格，也同时具有较高的食疗价值。荷花有清心凉血、清热解毒、美容养颜等作用，配合酸梅酱更能开胃消食、生津止渴，大暑时节宜赏宜食。

Ingredients

100g fresh lotus petals; 100g crispiness batter; 1 saucer of ground chili for dipping; 20g honey; 10g sour plum jam; 1 lotus seedpod; 1,000ml cooking oil (about 30ml to be consumed)

Preparation

1. Rinse the lotus petals in salty water, and drain. Mix the honey and plum jam in a bowl, blend well and transfer to a saucer.
2. Heat oil in a wok to 150℃, coat the petals with crispiness batter, and deep fry till golden brown. Transfer to a serving dish, and garnish with the lotus seedpod and lotus flower. Serve with the honey-plum saucer and chili saucer.

Features

golden in color; meltingly crispy in texture; beautiful presentation; mixed tastes of saltiness, spiciness, sourness and sweetness; strong lotus fragrance

Innovation

Lotus is a summer flower. The Greater Heat blunt people's appetite, but pink elegant lotus flowers help to lift the mood. Meanwhile, the Lotus Flower Festival in Danshan, Ziyang will launch lotus dishes to arouse people's appetite. This dish, using lotus petals as the main ingredients and decorated with lotus flowers and seedpods, has a high value in terms of art and food therapy. Lotus flowers which clear up the heart and cool down the blood are an ideal antidote to inner heat and a perfect food for beauty and skin. Sour plum jam promotes digestion, enhances the appetite, improves saliva production and slakes summer thirst. This dish is appealling to both the eye and the taste buds during the Greater Heat.

食材配方

天池原味藕粉100克　白糖20克　鱼胶片4张　清水800毫升　蜜饯50克　蜂蜜100克

制作工艺

① 蜜饯切成绿豆大小的颗粒；鱼胶放入清水中浸泡至软；藕粉和白糖混合均匀，加入清水调成稀糊状。

② 将鱼胶入沸水中煮化，先加入藕粉糊搅拌成透明的稠糊，再加入蜜饯颗粒拌匀，然后倒入方盘冷却后放冰箱中冷冻。

③ 将藕粉冻倒出，用刀或模具加工成多边形的块，装盘后淋上蜂蜜即可。

风味特色

色泽呈半透明状，质地细腻滑润，口味甘甜香醇。

创意设计

天池藕粉是乐至特产和国家地理标志产品。此品以天池藕粉为食材，采用冻的方法创制而成，老少皆宜。在养生方面，大暑时节暑湿交杂，出汗过多易伤阴耗气，藕粉能滋阴润肺、养心安神、清热生津，最宜此时食用，再配以蜂蜜，更添益气和胃之功。此外，藕粉所含淀粉为抗性淀粉，能避免血糖过快上升。

Ingredients

100g Tianchi lotus root powder (plain flavor); 20g sugar; 4 pieces of fish maw; 800ml water; 50g candied fruits; 100g lotus honey

Preparation

1. Cut the candied fruits into mung bean size grains. Soak the fish maw in water till soft. Mix the lotus root powder with sugar, add water, and stir well to make a paste.
2. Add the fish maw into boiling water, and boil until it melts. Add the paste, and continue to stir till transparent. Blend in the candied fruits, stir well, and pour into a square plate. Leave in room temperature to cool and then refrigerate.
3. Remove the jelly from the plate. Use a knife or mold to process the jelly into different shapes, transfer to a serving dish and drizzle with the honey.

Features

semi-transparent in appearance; soft and smooth texture; delicate sweet taste.

Innovation

Tianchi lotus root powder is a specialty of Lezhi County and a product of national geographical indications. This dish, using this specialty as the main ingredient and made by refrigerating, is preferred by people old and young. In terms of food therapy, during the Geater Heat when heat and humidity cause excessive sweating and harm qi circulation inside human body, lotus root powder is the best food, for it helps to relieve the lung, nourish the heart, calm the nerves, remove inner heat and promote saliva production. Its health benefits are doubled with honey, promoting qi circulation and soothing the stomach. Besides, the resistant starchy contents of lotus root powder reduce the rate of blood sugar rise.

THE BEGINNING OF AUTUMN

立秋

池水渐凉 金粟初开

立秋二绝（其一）

宋·范成大

折枝楸叶起园瓜，
赤小如珠咽井花。
洗濯烦襟酬节物，
安排笑口问生涯。

立秋物候 | PHENOLOGY IN THE BEGINNING OF AUTUMN

立秋是二十四节气中的第十三个节气，也是秋季的第一个节气。当太阳到达黄经135° 时为立秋，时间在每年的8月7日～9日。元朝吴澄《月令七十二候集解》言："立秋，七月节"，"秋，揫也，物于此而揫敛也"。秋，春华秋实，是植物快成熟的意思。立秋一般预示着炎热的夏天即将过去，秋天即将来临，草木开始结果，时序进入收获季节。

The Beginning of Autumn is the thirteenth of the 24 Solar Terms, and the first solar term in Autumn. It begins when the sun reaches the celestial longitude of 135°. It usually begins from 7th to 9th August every year. In his book *A Collective Interpretation of the Seventy-Two Phenological Terms*, Wu Cheng (Yuan Dynasty 1271-1368) said that the Beginning of Autumn is the July (lunar calendar) Festival. Qiu (Autumn) means that the plants are going to ripe. Normally, the Beginning of Autumn shows that the hot summer will go away, autumn will come soon, plants start to bear their fruits and it is the harvest season now.

我国古代将立秋后分为三候：一候凉风至；二候白露生；三候寒蝉鸣。意思是说立秋过后，刮的风会使人感觉凉爽；早晨的时候，大地上会有雾气产生；接着，秋天感阴而鸣的寒蝉也开始鸣叫。由于全国各地存在地理位置上的差异，地表及海拔的差异较大，南北东西不会在立秋这一天同时进入秋天。如在四川，立秋之后，常常仍然暑气难消，有时还会出现"秋老虎"的酷热，但总的趋势是天气逐渐转趋凉爽。

Ancient China has divided the 15 days after the Beginning of Autumn into 3 pentads: in the first pentad, the wind will make people feel cool; in the second pentad, there will be morning fog; in the third pentad, cicada in cold weather starts to chirp. Different areas will not enter Autumn together on the day of the Beginning of Autumn, because there are differences in geographical location, the earth's surface and altitude. For example, Sichuan is still hot after the Beginning of Autumn. There will be a spell of hot weather still. However, the general trend is getting colder and colder.

食材生产 | PRODUCING FOOD INGREDIENTS

立秋之后，在四川乃至西南地区，人们必须加强大秋作物的田间管理，促进其早熟，以避免受到低温霜冻的危害。据一些方志记载，"立秋宜昼。谚云：睁眼秋，收又收；闭眼秋，丢又丢。值小雨吉。""睁眼秋"的意思是说，如果上午立秋，根据民间经验，"秋老虎"不会太厉害，农作物的收成会比较好；"闭眼秋"的意思是说，如果下午立秋，"秋老虎"就比较猛，可能会减少收获。这一时节，如果能够遇小雨，则更有利于农业生产。秋季是丰收的季节，不仅稻谷飘香，自立秋时节开始，还有许多蔬菜瓜果进入成熟采摘期，资阳地区著名的安岳紫薯就是其中之一。

After the Beginning of Autumn, people must strengthen fields management to promote maturing and prevent frost damage in Sichuan and South west China. According to some local chronicles, the spell of hot weather after the Beginning of Autumn won't be severe and the harvest will be good if the Beginning of Autumn starts in morning; the spell of hot weather after the Beginning of Autumn will be severe and the harvest will be bad if it starts in afternoon. If it is rainy in this solar term, agricultural production will be fine.

Autumn is the season of harvest. Since the Beginning of Autumn, not only rice, numerous kinds of vegetables and fruits are ripe. Among them, there is purple sweet potato from Anyue, Ziyang City.

为迎接秋天、酬谢秋收，我国古代在立秋时节也有很多相关的民俗活动。周朝时，每逢立秋，周王会亲自率领大臣到西郊迎秋，举行祭祀、蓐收等仪式。到宋朝，立秋之日，皇帝会在宫中举行迎秋仪式。而在民间，百姓则在立秋后的第五个戊日举行祭祀土地神的仪式，以酬谢丰收，这就是“秋社”日，这一传统习俗一直延续至今日。

In order to welcome autumn, there were lots of folk activities in the Beginning of Autumn in ancient China. In Zhou Dynasty, the kings would hold ritual to welcome Autumn with all his courtiers in western suburbs. In Song Dynasty, emperors would hold ceremony to welcome autumn in palace. Common people will hold a ceremony for the god of land on the fifth Xu day after the Beginning of Autumn till today.

饮食养生 | DIETARY REGIMEN

立秋开始，天气慢慢转凉，早、晚与中午开始出现温差。虽然早、晚天气逐渐变凉，但白天往往盛夏余热未消，秋阳肆虐，四川许多地方还常常处于炎热之中，故素有“秋老虎”之称。在立秋时节，不仅要清热生津，以应对余热未消的“秋老虎”，还要增强体质、预防感冒等各种呼吸道疾病的侵袭。许多人为了补充夏季人体的损失，在立秋后开始“贴秋膘”，即多吃肉及味厚的美食。但需要注意的是，当今的人们往往食用肉类、脂肪较多，而体力劳动、运动不足，容易出现血糖高、血脂高、血压高的“三高”现象，因此“贴秋膘”也应适度，因人而异。我国传统医学认为，秋季对应于肺，肺为娇脏，喜欢温暖、湿润，最容易被燥气所伤，立秋之后雨水减少、气候渐燥，不仅易伤肠津而导致便秘，也对肺会产生较大损伤。所以，秋季应格外注重养肺，饮食应滋阴润燥，如多吃白萝卜、白菜、莴笋、油菜、菠菜、茭白、苹果、香蕉等蔬果以润肠通便，也可食用银耳、莲藕、梨、蜂蜜、荸荠润肺滋阴，还可适量食用百合、玉竹、麦冬、白果等养肺润燥之品。

After the Beginning of Autumn, it is getting cool and temperature difference appears in the morning, in the evening and at noon. Although it is getting cool in the morning and evening, it is still hot during day time. Many parts of Sichuan are still hot because of the spell of hot weather. In the Beginning of Autumn, people need clear the summer heat to deal with the spell of hot weather, and need strengthen physique to prevent influenza and other respiratory diseases. In order to supplement human body's loss in summer, many people choose to eat more meat and thick taste foods. However, it is worth noting that people have suffered from Three Highs (hyperlipidemia, hyperglycemia, hypertension) by eating more meat and fat, and do less physical labor and sports nowadays. Thus, eating in autumn should be properly and based on individual situations. TCM believes that in autumn we should take care of the lung. Lung as a delicate organ prefers to warm and humid condition, and is easy to be affected by dryness. Rain fall is getting less and climate is getting dry after the Beginning of Autumn, which will affect lung and intestinal tract and even bring about constipation. Therefore, people should pay special attention to tonify lung in autumn by eating white radish, cabbage, lettuce, rape, spinach, cane shoots, apple, banana which are good for intestinal tract; and nourishing lung by eating tremella, lotus root, pearl, honey, chufa, lily, polygonatum odoratum, radix ophiopogonis, ginkgo.

立秋美食 | FINE FOODS IN THE BEGINNING OF AUTUMN

立秋时节，各种瓜果、蔬菜大量成熟，种类很多，鱼肉禽蛋也比较丰富，民间出现了较多饮食习俗，除了“贴秋膘”，还有“尝秋鲜”“咬秋”等。所谓“尝秋鲜”，是指品尝秋天收获的粮食和各种瓜果、蔬菜，人们认为食用新鲜出产的粮食、蔬果，不仅味道最美，而且最富营养；所谓“咬秋”也称“啃秋”，是指在立秋这天吃西瓜或香瓜，不仅要“咬”住秋天带来的凉爽，更是在辛勤劳作后“咬”住丰收的喜悦。此外，民间在立秋时开展“秋社”这一祭祀活动时，也会在家中准备各种美食，款待宾客。

In the Beginning of Autumn, all kinds of vegetables, melons and fruits are ripe. There are lots fish and eggs on the market too. There are lots of eating customs, like gain on weight in autumn, taste what is just in autumn, biting autumn, etc. Tasting what is just in autumn means trying all kinds of melons, fruits and vegetables harvested in autumn. People believe that newly harvested food is most delicious and full of nutrition. Biting autumn means eating watermelon or muskmelon on the day of the Beginning of Autumn. People want to enjoy not only the coolness of autumn, but also the pleasure after good harvest. Moreover, people will hold the ritual activity called “Qiu She” in the Beginning of Autumn, and prepare all kinds of delicious food to host the guest.

立秋节气的时间，在不少年份常与农历七月七的七夕节相近或相遇。在七夕这一天，民间有制作“巧果”以乞巧的习俗。巧果主要是用面粉、米粉、豆面、薯面等做成的花式糕点，其形态有飞禽走兽或花果玩物，多栩栩如生。在四川民间还有做“巧芽”乞巧的习俗，即在七夕之前用水浸豌豆于碗中，“令芽长尺余，红线束之，名曰‘巧芽’”，“摘取芽尖投水中，对灯、月下照之，或现针影，或露花影，或变鱼、龙影，相与为欢，谓之‘得巧’”。

The date of the Beginning of Autumn is close to the Chinese Valentine’s Day in some years. On the Chinese Valentine’s Day, people will make some kind of beautiful cake which made of flour, rice flour, bean flour, sweet potato flour. People can give it the vivid shape of birds, animals, flowers or fruits. There is a folk custom of steeping peas in bowls and make them sprout. On the Chinese Valentine’s Day, people bind the roots with red threads, cut the roots and put them in water. With the lamp light or moon light, people can watch the shadows of roots, like needles, flowers, fishes or dragon.

锅魁坛子肉
Tanzi Pork with Guokui

食材配方

坛子肉300克　小锅魁5个　小黄瓜300克　泡仔姜100克　小番茄10个　生菜50克　甜面酱60克　白糖10克　酱油5毫升　芝麻油3毫升　三色堇两朵

制作工艺

① 将甜面酱、白糖、酱油、芝麻油调匀，装入用小黄瓜雕刻成的小盏中成味碟。

② 泡仔姜切成细丝；小黄瓜斜切成片。

③ 坛子肉入笼蒸熟，再切成长约8厘米、宽约4厘米、厚约0.2厘米的片。

④ 将小黄瓜片分别摆放在10个盘子的右上方，每个盘中放4片坛子肉，右下方放味碟，左上方放夹有生菜的锅魁1/2个，点缀上小番茄、仔姜丝、三色堇即成。

风味特色

色彩分明，味道咸鲜，坛子肉的醇香突出，口感丰富。

创意设计

坛子肉是资阳市著名的加工性特产食材，境内各区县都有制作和在菜点中运用，营养丰富、风味独具一格。此菜采用旱蒸法制成，荤素搭配、色彩分明、味道咸鲜、开胃不腻。在养生方面，猪肉养阴润燥，是立秋“贴秋膘”的首选；仔姜辛温发散、开胃生津；黄瓜、番茄、生菜滋阴清热、减少油腻，还能制约仔姜的温热之性，适宜此时湿热的气候。

Ingredients

300g Tanzi Pork (pork in a crock); 5 small guokui (flatbread); 300g small cucumbers; 100g tender ginger, pickled; 10 small tomatoes; 50g lettuce; 60g fermented flour paste; 10g sugar; 5ml soy sauce; 3ml sesame oil; 2 pansies

Preparation

1. Mix the fermented flour paste, sugar, soy sauce and sesame oil to make the dipping sauce. Carve the small cucumbers into 10 saucers. Transfer the sauce to cucumber saucers.
2. Shred the pickled ginger. Slice small cucumbers diagonally.
3. Steam the pork in a steamer till cooked through, and cut into 8cm long, 4cm wide and 0.2cm thick slices.
4. Lay the cucumber slices on 10 serving plates' upper right-hand side. Place four pork slices on each plate, a saucer at its lower right-hand side, and half a guokui stuffed with lettuce at its upper left-hand. Garnish with small tomatoes, ginger slivers and pansies.

Features

distinct colors; savory taste; lingering and rich flavors

Notes

Tanzi Pork, or Pork in a Crock if translated literally, is a famous processed food in Ziyang City, which is widely used in every district in this area. This ingredient is not only rich in nutrition, but also has unique flavors. With steamed pork and vegetables, this dish has distinct colors, savory and non-greasy taste. In terms of food therapy, pork has the nature of nourishing yin and moisturizing dryness of our bodies. During the Beginning of Autumn, it is a perfect choice for increasing body fat, which will help defend the coming cold winter. Tender ginger is pungent and can increase the appetite and stimulate saliva production. Cucumbers, tomatoes and lettuce can nourish yin and dispel inner heat, reduce greasiness and balance the warm nature of ginger, all suitable for the humid and warm weather during this period.

爽口巧芽

Appetizing Qiaoya

食材配方

花生芽100克　黄豆芽100克　藕带100克　小米辣椒20克　野山椒50克　泡菜水1000毫升　食盐10克

制作工艺

① 花生芽、黄豆芽去根后洗净；藕带刮去外皮，洗净后切成长约4厘米的节，入清水中浸泡待用。

② 小米辣椒、野山椒分别切为颗粒后入坛，再在坛中注入用泡菜水、食盐调成的泡菜盐水。

③ 将花生芽、黄豆芽、藕带分别焯水后捞出，先放入凉开水中浸泡至冷，再入泡菜坛中浸泡4小时后捞出装盘即成。

风味特色

晶莹透亮，味道咸酸微辣，质地嫩脆爽口。

创意设计

所谓“巧芽”，是指用五谷杂粮的种子经浸泡而生长出的嫩芽。立秋节气常与“七夕”相近，四川民间在“七夕”时曾有做巧芽、赏巧芽，以巧芽祭月等习俗，此菜的创意即来源于此。在养生方面，“巧芽”经泡制发酵，有利于大量维生素的保存和钙、铁等矿物质的吸收，泡野山椒酸辣开胃，还能增进食欲。

Ingredients

100g peanut sprouts; 100g soybean sprouts; 100g lotus root sprouts; 20g bird's eye chilies; 50g mountain chilies; 1,000g pickle brine; 10g salt

Preparation

1. Remove the roots of peanut sprouts and soybean sprouts, and rinse. Peel lotus root sprouts, rinse, cut into 4cm long segments, and soak in water.
2. Chop bird's eye chilies and mountain chilies into cubes and put into a crock. Mix the brine with salt and pour into the crock.
3. Blanch peanut sprouts, soybean sprouts and lotus root sprouts, transfer to cold boiled water to cool. Add to the crock and pickle for 4 hours before serving.

Features

translucent looking; sour, spicy and savory flavors; fresh, crispy and tender tastes

Notes

The Beginning of Autumn is close to Chinese Valentine's Day, the 7th day of July in lunar calendar. Local people have the customs of preparing Qiaoya, appreciating Qiaoya, and offering Qiaoya to the moon on this occasion. As for health benefits, pickling and fermenting help to preserve vegetable vitamins, and promote absorption of minerals like iron, calcium and so on. Meanwhile, pickled mountain chilies can stimulate the appetite.

安岳紫薯包

Anyue Purple Yam Baozi

食材配方

面粉200克　酵母2克　泡打粉1克　砂糖4克　紫薯粉12克　清水100毫升
猪油8克　豆沙馅心250克

制作工艺

① 将面粉、酵母、泡打粉、砂糖、紫薯粉混合均匀，加清水调制成面团，再加入猪油反复揉搓至面团光滑，醒发备用。

② 把面团分成重约20克一个的剂子，包入5克豆沙馅制成紫薯包，再放入醒发箱醒发15分钟。

③ 将醒发好的紫薯包放入蒸锅，用大火蒸制15分钟后取出，趁热且无水汽时在表面粘上紫薯粉即可。

风味特色

造型逼真，味道甜香，质地松软。

创意设计

紫薯营养丰富，加入面团中，再包上豆沙馅，做成象形紫薯包，口感香甜松软，外观逼真，饶有情趣。在养生方面，赤小豆馅心能祛湿利水，辅以面粉健脾温中、紫薯益气资肾，可谓祛湿健脾佳点。此外，紫薯富含膳食纤维、维生素、微量元素、花青素、糖蛋白等有利于人体健康的功能因子，具有抗氧化、抗肿瘤等作用。

Ingredients

200g wheat flour; 2g yeast; 1g baking powder; 4g sugar; 12g purple yam powder; 100ml water; 8g lard; 250g mashed red beans

Preparation

1. Mix the wheat flour, yeast, baking powder, sugar and purple yam powder, and add water to make dough. Add lard and knead till the dough is smooth. Set aside to proof.
2. Cut the dough into 20g equal portions. Use each portion to wrap up 5g mashed red beans, and shape into purple yams. Proof in a proofer for 15 minutes.
3. Steam the stuffed dough in a steamer over a high flame for 15 minutes, remove, and coat with a thin layer of purple yam powder while still hot without vapor. Serve the yam baozi.

Features

vivid shape; sweet taste; soft texture

Notes

This yam shaped dish has sweet taste and soft texture. In food therapy, red beans can dispel inner dampness and promote body water excretion. Wheat flour can warm the body and nourish the spleen. Purple yam can nourish the kidney and the spleen, strengthen qi and dispel dampness. Meanwhile, purple yam has antioxidant and anti-tumor functions because it is rich in dietary fiber, vitamins, minerals, anthocyanin and glycoprotein, etc.

THE END OF HEAT

长江二首（其一）

宋·苏泂

处暑无三日，
新凉直万金。
白头更世事，
青草印禅心。
放鹤婆娑舞，
听蛩断续吟。
极知仁者寿，
未必海之深。

处暑

乾坤渐肃　残暑除空

处暑物候 | PHENOLOGY IN THE END OF HEAT

处暑是二十四节气中的第十四个节气。当太阳到达黄经150° 时为处暑，时间在每年的8月22日～24日。元朝吴澄《月令七十二候集解》说："处，去也，暑气至此而止矣。" "处"是终止的意思，表示暑气、炎热即将过去，我国大部分地区气温逐渐下降。处暑既不同于小暑、大暑，也不同于小寒、大寒，它是反映气温变化的一个节气，代表气温由炎热向寒冷过渡。处暑之后，暑气虽然逐渐消退，但是炎热并不会立即消失。

The End of Heat is the fourteenth of the 24 Solar Terms. It begins when the sun reaches the celestial longitude of 150°. It usually begins from 22nd to 24th August every year. In his book *A Collective Interpretation of the Seventy-Two Phenological Terms*, Wu Cheng (Yuan Dynasty 1271-1368) said that the End of Heat in Chinese is "Chu Shu", and the word "Chu" means the end of summer heat. In most parts of China, temperature is getting down gradually. The End of Heat is different from Lesser Heat, Greater Heat, Lesser Cold and Greater Cold. It reflects the change of temperature, and means that the weather goes from hot towards cold. After the End of Heat, summer heat is fading away, but the hotness won't disappear immediately.

我国古代把处暑后的十五天分为三候：一候鹰乃祭鸟；二候天地始肃；三候禾乃登。意为在天气由热转凉的处暑时节，老鹰开始大量捕食鸟类；天地万物开始凋零；农作物在此时开始成熟并收获。

Ancient China has divided the 15 days after the End of Heat into 3 pentads: in the first pentad, eagles begin to hunt other birds; in the second pentad, plants start to wither; in the third pentad, crops are ripe and harvested.

食材生产 | PRODUCING FOOD INGREDIENTS

处暑之后，四川地区的平均气温可降到23℃～27℃，雨量大多在150～200毫米。中稻普遍成熟，收打正忙，其他部分秋收作物也进入了收获期。在资阳，雁江蜜柑在此时逐渐开始成熟上市，比同类型气候条件下的重庆、湖北、湖南、陕西、贵州等地产的蜜柑要早成熟上市15～25天。同时，四川也有部分作物在处暑之后进行种植。清朝乾隆年间，四川人张宗法结合当时四川的农业特色撰写了综合性农学著作《三农纪》。书中记载了许多节气的农业生产状况，言处暑时节前后，宜"植葱、植大蒜、种萝卜、伐竹木、栽藠蒜、种诸菘、种诸芥菜、种秋荞、植莴苣、植菠菜"。

After the End of Heat, the average temperature of Sichuan region is from 23℃to 27℃ and the rain fall is from 150mm to 200mm. Mid-season rice is ripe generally. Some kharif crops are in harvest season. In Ziyang City, the tangerine from Yanjiang is on the market. It is 15 days to 25 days earlier than those planted in Chongqing, Hubei, Hunan, Shaanxi, Guizhou which have the same climate as Ziyang City. Meanwhile, people will plant some other crops after the End of Heat. In Qianlong Period, Qing Dynasty, Zhang Zongfa from Sichuan had written a comprehensive agriculture work *San Nong Ji* on the base of Sichuan's agriculture experiences. This book has recorded the agricultural production in many solar terms. It said that green onion, garlic, radish, bamboo, allium chinense, chinese cabbage, mustard leaf, lettuce and spinach are suitable to plant around the End of Heat.

饮食养生 | DIETARY REGIMEN

处暑之后的天气特点是白天热、早晚凉，昼夜温差大，降雨减少，空气湿度降低。在这样的环境下，人体容易出现口鼻干燥、咽干唇干的肺燥证。此时，阳气渐收，人们需要保护好阳气，除了减少空调、电扇的使用外，还需要从饮食着手进行调养。处暑前后，许多鱼虾及贝类发育成熟、肉质肥美，但河鲜类多属寒凉性质，不宜多食，且食用时宜搭配姜、蒜等热性食物。若感觉口干舌燥，甚至困倦疲劳，除多吃滋阴润肺的食物外，还可选择西洋参来补气养阴；若睡眠不佳，则可食用莲子以养心安神。此外，莲子、芡实都具有收敛固涩的作用，与鸭肉等一同煲汤，秋季正宜。

The features of climate after the End of Heat are hot in the daytime, cool in the morning and evening, and there is temperature difference between day and night, less rainfall and low humidity. In that circumstance, human body will suffer from dryness of lung, which reflects in dry mouth, dry nose and lips. In this period, yang qi in human body is not strong as before, hence, people should protect yang qi by reducing usage of air conditioning and electric fan, and take proper foods. Around the End of Heat, fishes, shrimps and shellfishes are mature and taste good. Because those food ingredients from river are "cold food", people can not eat too much and they need be eaten with "hot food" like ginger and garlic. If mouth and tongue are dry, or body feels drowsy, people should eat more food which can tonify the lung, or try American ginseng which can tonify energy. If you could not sleep well, you can eat lotus seeds to calm the nerves. Moreover, louts seed and gorgon fruit have special function in autumn, and should be made into soup with duck.

处暑美食 | FINE FOODS IN THE END OF HEAT

处暑时节，一些地区有吃鸭子的传统。鸭子味甘性凉，能够滋养肺、胃，对消除口舌干渴、疲乏无力等症有一定作用。在资阳，鸭子的制法极多，著名品种有仔姜鸭等。此外，处暑节气前后的民俗多与祭祖及迎秋有关。处暑的时间，在一些年份与农历七月十五的“中元节”相近。中元节又称作“七月半”，在四川许多地区，民间还有“具酒馔，荐祭祖宗”的习俗。

There are traditions of eating ducks in some region in the End of Heat. Duck meat is cool in nature, which can tonify lung and stomach, and be helpful to dry mouth and fatigue. There are many ways of cooking ducks in Ziyang City, like famous Ginger Duck. Moreover, the folk customs around the End of Heat are connected with worshiping ancestors and welcoming autumn. The date of the End of Heat is close to the Ghost Festival (Zhongyuan Festival). It is called Mid-July as well. In many parts of Sichuan, there are tradition of worshiping ancestors with liquor and dishes.

苌弘鲶鱼

Changhong Catfish

食材配方

鲶鱼1尾（约1500克） 仔姜100克 郫县豆瓣50克 小米辣50克 姜米6克 蒜米10克 泡辣椒段30克 香葱段20克 芹菜节50克 啤酒100毫升 味精3克 鲜汤1000毫升 食用油100毫升

制作工艺

① 鲶鱼宰杀、洗净后斩成条；仔姜切成二粗丝。

② 锅置火上，入油烧至120℃时，放入郫县豆瓣、小米辣、姜米、蒜米、泡辣椒段、仔姜炒香至油呈红色，掺入鲜汤，入鲶鱼、啤酒烧至成熟，再入香葱段、芹菜节、味精烧沸，起锅装盘即成。

风味特色

色泽红亮，肉质细嫩，味道咸鲜香辣。

创意设计

此菜的创意源于苌弘与孔子一起品尝鱼肴的故事，是资阳著名的特色菜肴，历史悠久。在养生方面，鲶鱼滋阴养血、利水消肿。虽然郫县豆瓣、小米辣辛辣发散，但啤酒能清热养阴，对辛散有所制约，更使此菜食疗价值显著。此外，鲶鱼是高蛋白、低脂肪的代表，营养价值较高。

Ingredients

1 catfish (about 1,500g); 100g tender ginger; 50g Pixian chili bean paste; 50g bird's eye chilies; 6g ginger, finely chopped; 10g garlic, finely chopped; 30g pickled chilies, segmented; 20g spring onions, segmented; 50g celery, segmented; 100ml beer; 3g MSG; 1,000ml stock; 100ml cooking oil

Preparation

1. Cut the fish into strips. Shred the ginger into medium slivers.
2. Heat oil in a wok to 120℃, add Pixian chili bean paste, bird's eye chilies, ginger, garlic, pickled chilies, ginger, and stir fry till aromatic and lustrous. Pour in the stock, add the fish and beer, and braise till cooked through. Add spring onions, celery and MSG, bring to a boil, and transfer to a serving dish.

Features

reddish brown color; tender fish; savory and spicy tastes

Notes

This dish originates from a story related to Confucius and Changhong while they were having a fish meal together. It is a famous, time-honored Ziyang dish. In food therapy, catfish is good for nourishing yin and blood, promoting body water excretion and relieving edema. Although Pixian chili bean paste and bird's eye chilies are pungent, beer can balance them, for it helps dispel inner heat and nourish yin. Besides, catfish is nutritious with its high protein and low fat.

苦藠鸭掌 | Kujiao Duck Feet

食材配方

鸭掌750克　苦藠250克　猪肚250克　姜片5克　葱段3克　红枣5克　枸杞2克　食盐12克　味精3克　猪骨汤2000毫升

制作工艺

① 猪肚焯水后切成条；鸭掌焯水后备用。

② 砂锅置火上，放入鸭掌、肚条、姜片、葱段、红枣、枸杞、猪骨汤、苦藠，置旺火上烧沸后，改用小火煨至鸭肉软熟，最后入食盐、味精调味即成。

风味特色

鸭掌质地软糯，苦藠微苦，汤汁咸鲜清爽。

创意设计

此菜鸭掌软糯、苦藠微苦，汤清淡爽口，深得百姓喜爱。在养生方面，鸭掌滋阴利水、苦藠清热健胃、猪肚健脾补中，三者结合可共收养阴益气之效，尤宜秋季滋补。此外，鸭掌富含胶原蛋白，有美颜嫩肤等作用。

Ingredients

750g duck feet; 250g kujiao bulbs; 250g pork tripe; 5g ginger, sliced; 3g spring onions, segmented; 5g red dates; 2g wolfberries; 12g salt; 3g MSG; 2,000ml pork bone stock

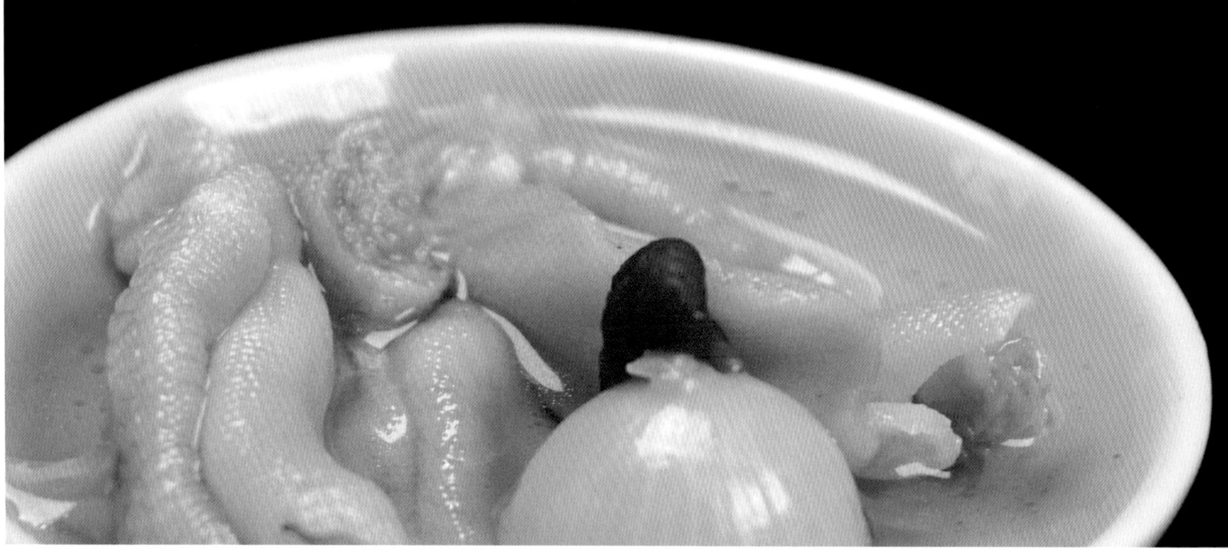

Preparation

1. Blanch the pork tripe and slice into strips. Blanch the duck feet.
2. Place an earthen pot over the flame, add the duck feet, tripe stripes, ginger, spring onions, red dates, wolfberries, stock and kujiao bulbs, bring to a boil over a high flame, and simmer over a low flame till duck is soft and cooked through. Add salt and MSG to season.

Features

tender and glutinous duck feet; slightly bitter kujiao bulbs; savory and fresh soup

Notes

In this dish, duck feet are soft and glutinous, while kujiao is bitter. However, because of this, this soup is fresh and light, which is widely welcomed by people. In food therapy, duck feet can nourish yin and promote body water excretion. Kujiao can dispel inner heat and nourish the stomach. Pork tripe can nourish the spleen. Combining these three ingredients, this dish replenishes qi and nourishes yin, especially suitable for autumn. Additionally, duck feet are rich in collagen, which can beautify skin.

柠檬酥 Lemon Pies

食材配方

面粉250克　猪油60克　黄油60克　清水100毫升　吉士粉2克　鸡蛋黄1个
食用油1000毫升（实耗50毫升）　柠檬1个（榨汁）　莲蓉馅150克　酥花生碎30克

制作工艺

①将面粉、猪油、黄油、吉士粉、鸡蛋黄、柠檬汁和清水按照比例分别调制成水油面团和油酥面团待用。

②将水油面包油酥面擀成长方形，对叠封口后擀薄，反复两次折三层，擀薄到0.4厘米厚，制作成叠酥，入冰箱冷藏待用。

③莲蓉馅里加入酥花生碎，拌制成柠檬馅心。

④将开好酥的面团切成薄片，取面皮包入柠檬馅心，收紧封口后捏成柠檬形生坯。

⑤锅置火上，入食用油烧至90℃，放入柠檬形生坯，用中火炸至酥层清晰、层次变硬时起锅装盘即成。

风味特色

形色似柠檬，口感化渣酥松，酥层分明清晰，柠檬香味突出。

创意设计

柠檬酥是利用传统明酥开酥方式创新设计的象形酥点。在养生方面，莲子养心安神、健脾益肾，加糖和油脂为馅更为滋润养阴。此外，莲子的蛋白质、碳水化合物丰富，有助于防癌抗癌、降低血压。

Ingredients

250g wheat flour; 60g lard; 60g butter; 100ml water; 2g custard powder; 1 egg yolk; 1,000ml cooking oil (50ml to be consumed); 1 lemon (juiced); 150g lotus seed paste; 30g crispy peanuts, crushed

Preparation

1. Make lard dough and short dough with the flour, lard, butter, custard powder, egg yolk, lemon juice and water.
2. Wrap up the short dough with the lard dough and roll it into the shape of a rectangle. Fold in half, seal the edges, and flatten with the rolling pin. Repeat the above folding and flattening process two more times. Fold the dough into three layers, roll into a 0.4 cm thick piece, and transfer into a refrigerator.
3. Mix the crushed crispy peanuts with the lotus seed paste to make the stuffing of the pies.
4. Cut the dough into thin wrappers, wrap up the stuffing, seal and shape into lemons.
5. Heat up the cooking oil to 90℃ in a wok, add the lemon-shaped dough, and deep fry over a medium flame until the pies are cooked through with hard and clear layers.

Features

lemon shaped flaky pies with strong lemon fragrance; meltingly crispy texture

Notes

Lemon pie is a creative, pictographic snack using the traditional cooking method of making flaky pastry. In food therapy, lotus seeds can nourish the heart, tranquilize the mind, strengthen the spleen and tonify the kidney. The pie has a stronger function of nourishing yin with sugar and oil. In addition, the rich protein and carbohydrate contained in lotus seeds help to prevent and fight cancer and reduce blood pressure.

WHITE DEW

白露

唐·杜甫

白露团甘子，
清晨散马蹄。
圃开连石树，
船渡入江溪。
凭几看鱼乐，
回鞭急鸟栖。
渐知秋实美，
幽径恐多蹊。

白露

玉露生凉　草木凝烟

白露物候 | PHENOLOGY IN WHITE DEW

白露是二十四节气中的第十五个节气。当太阳到达黄经165° 时为白露，时间在每年的9月7日～9日。露是水汽因为气温降低而在地面或近地物体上凝结形成的水珠。《礼记》说："凉风至，白露降，寒蝉鸣。"元朝吴澄《月令七十二候集解》言："白露，八月节""阴气渐重，露凝而白也"。白露表明天气继续转凉。

White Dew is the fifteenth of the 24 Solar Terms. It begins when the sun reaches the celestial longitude of 165°. It usually begins from 7th to 9th September every year. Dew is the very small drops of water that form on the ground because the low temperature. *The Book of Rites* said that when the cool wind comes, there is dew and the winter cicada begins to chirp. In his book *A Collective Interpretation of the Seventy-Two Phenological Terms*, Wu Cheng (Yuan Dynasty 1271-1368) said that White Dew is the August (lunar calendar) Festival. It is getting colder and dew appears.

我国古代把白露后的十五天分为三候：一候鸿雁来；二候玄鸟归；三候群鸟养羞。意思是说，随着天气不断转凉，白露后的第一个五日，大雁从北方飞来；再过五日，燕子向南归去；又经五日，群鸟就开始贮藏过冬的食物了。

Ancient China has divided the 15 days after White Dew into 3 pentads: in the first pentad, wild geese have flown from the north; in the second pentad, swallows will fly to the south; in the third pentad, birds will store foods for winter.

食材生产 | PRODUCING FOOD INGREDIENTS

白露时节，包括四川在内的西南地区常常处于收割稻谷的重要阶段。但是，在四川盆地，白露后的日平均气温可降到17℃～22℃，降雨量约70～120毫米，开始进入秋雨绵绵的季节。因此，四川地方志中常言，白露"是日不宜雨，稻沾之粘（粒）秕，蔬沾之味苦"。这说明白露时节最不宜降雨，否则对稻米和蔬菜的成熟、收获极为不利。在资阳，此时除了开始收割稻谷，当地特产的柠檬、青花椒、黄花菜等食材也纷纷成熟上市。如安岳柠檬，年产量约占全国总产量的70%，鲜果及其加工产品销往国内外。此外，这一节气还是秋冬甘蓝、中晚熟花椰菜、小白菜、韭黄、水芹菜等食材的重要种植时期。

In White Dew, the Southwest China including Sichuan is in the important stage of harvesting rice. However, the average temperature after White Dew is from 17℃ to 22℃, and the average rain fall is from 70 mm to 120 mm. It is the persistent autumn rain time. In Sichuan local chronicles, raining in White Dew is not a good news, which will hinder the ripening and harvest of grain and vegetable. In Ziyang City, local lemon, green Sichuan pepper and day lily have gone on the market, in addition to harvesting rice. Fresh or processed lemon from Anyue County, whose annual output accounts for 70% of national production, have been sold on domestic and international markets. Furthermore, White Dew is also an important period to plant autumn and winter cabbage, medium and late cauliflower, pakchoi, chive, cress.

饮食养生 | DIETARY REGIMEN

白露是典型的秋天节气，气温渐凉，在饮食上也要更加慎重。《礼记》中针对季节变化明确提出了饮食与四时相应调和的观点："凡和，春多酸，夏多苦，秋多辛，冬多咸。"意思是说，春季时人体刚经过冬季，内部活跃度低，可适当吃一些酸性食物来促进胃动力，帮助消化吸收；夏季时天气湿热、人体易上火，可多吃寒凉性质的苦味食物来中和体内热气；秋季时天气转凉，人体阴气增加，可吃一些温热性质的辛味食物，以起到发散解表、行气活血等作用；冬季时天气寒冷，可适当吃一些咸味食物，起到"冬藏"的作用。在四川，大多从白露节气开始秋雨绵绵，少见太阳，人们则更加喜爱用辣椒、花椒等辛辣味食材制作菜肴，既能刺激味蕾、获得味觉之美，还能行气活血、祛风除湿，促进身体健康。此外，随着气温的降低，还可多吃一些增强体质的食品，如牛肉、猪肝、鸡、鸭、鱼、山药等。白果也是此时的佳品，具有益肺气、治咳喘、护血管等食疗作用，但需注意白果有微毒，需要完全做熟且控制好用量。

White Dew is a typical autumn solar term. People should be cautious when it is getting cold. *The Book of Rites* has pointed out that diet should be reconciled with seasons. In spring, the human body is not active because it just experiences winter; then people should eat some sour foods to motivate the stomach, help digestion and absorption. In summer, the body is too hot because of the weather; then people should eat bitter "cold" foods to clear the inner heat. In autumn, the yin qi in the body is getting strong; then people should eat warm pungent food to activate the blood and energy. In winter, it is cold and people should eat some salty foods. In Sichuan, it is rainy and overcast since White Dew. Hence, people here like to cook with chilly and Sichuan pepper, which can shake up the taste buds and activate the blood, clear the humidity, keep healthy. When the temperature drops, people can also eat some foods which can strengthen physique, like beef, pork liver, chicken, duck, fish, Chinese yam, etc. Ginkgo is the proper food in this period. It can tonify the lung, relieve cough and protect the blood vessel. However, we must control the amount of ginkgo and cook it well, because it is a little bit toxic.

白露美食 | FINE FOODS IN WHITE DEW

白露时节，全国各地都有许多饮食习俗，如在南方一些地区会喝白露茶、酿白露米酒，一些地方则采集"三样白"或"十样白"（白木槿、白毛苦等名字中带"白"的草药）与乌鸡或鸭子一同煨制。此外，白露节气的时间，在许多年份也常与农历八月十五的"中秋节"相近。在四川，不同的地方志中都记载了中秋的美食，不仅是月饼，还有其他品种。如1983年版温江地区《武阳镇志》载："各户置酒菜过节。晚间吃月饼、麻饼、麻糖、水果，合家欢聚。"在资阳市的乐至县，许多人更喜欢购买和食用当地的著名面点"乐至麻饼"。

In White Dew, there are many eating customs all over the country. In south, people will drink White Dew tea, brew White Dew wine. In some places, people will cook silkie or duck with 3 kinds or 10 kinds of Chinese herbs which has the word "white" in their names. The date of White Dew is close to the Mid-Autumn Festival in some years. Different local chronicles have recorded the fine foods in Mid-Autumn Festival, including moon cake and others. *Wuyang Town Chronicles* (1983 edition) recorded that all families prepared liquors and dishes for the festival. People stayed with families, eating moon cakes, sesame cakes, sesame candies, fruits during the evening. In Lezhi County, people prefer to purchase and eat local famous Lezhi Sesame Cake.

食材配方

鲜柠檬500克　白糖160克　清水160毫升　蜂蜜50毫升　食盐10克

制作工艺

①柠檬切片、去籽，用清水浸泡4次，去掉柠檬的苦涩味，捞出后沥干水分。

②锅置火上，入清水、白糖，用小火熬至浓稠，加入食盐后将锅端离火源，待糖水冷却到40℃左右时加入蜂蜜、柠檬片，再放在火上，用小火加热；如此反复3次，捞出装盘即成。

风味特色

色泽淡黄，晶莹透亮，味道甜香。

创意设计

柠檬是安岳的著名特产食材。此菜用蜜汁方法制成，色泽淡黄、晶莹透亮，味道香甜。在养生方面，酸味的柠檬辅以蜜汁，酸甘化阴，更能养阴生津、养肺润燥。此外，柠檬富含柠檬酸和多种维生素、微量元素，具有美容保健、预防高血压和心血管疾病等作用。

水晶柠檬 Crystal Lemon

Ingredients

500g lemons; 160g sugar; 160ml water; 50ml honey; 10g salt

Preparation

1. Cut the lemons into slices and remove the seeds. Soak the lemon slices in water to remove the bitter and harsh taste. The water needs to be changed four times during the process. Remove and drain.
2. Heat a wok, add water and sugar, simmer over a low flame till the sugar melts and the sauce becomes viscous and thick. Add the salt, remove from the flame, and leave to cool to about 40℃. Blend in the honey and lemon pieces, and reheat over a low flame. Repeat the above steps two more times and transfer to a serving dish.

Features

light yellow color; crystal clear appearance; sweet and fragrant tastes

Notes

This dish uses lemon, a local specialty of Anyue County, as the main ingredient. In food therapy, sour lemon, when combined with sweet sugar sauce, has a stronger function in nourishing yin, promoting saliva production, and relieving the lung. In addition, lemon is rich in citric acid, various vitamins and microelements, so it is not only good for beauty and health, but prevents high blood pressure and cardiovascular diseases.

食材配方

乳鸽2只　糯米30克　银杏果50克　火腿丁15克　火腿片30g　冬笋丁15g　冬笋片30g　水发香菇丁15g　水发香菇片30g　鲜汤500毫升　郫县豆瓣40克　食盐6克　料酒10毫升　味精1克　胡椒粉0.5克　芝麻油3毫升　水淀粉10克　姜米10克　葱花15克　食用油1000毫升（约耗70毫升）

制作工艺

① 糯米入锅煮至8成熟，捞出后加入食盐、银杏果、火腿丁、冬笋丁、水发香菇丁拌匀成馅料。

② 将乳鸽从颈背部进刀整料去骨，洗净后将馅料填入腹内，用牙签封住刀口，入沸水中焯水，用牙签在鸽子身上扎眼放气。

③ 锅置火上，入油烧至120℃时，放入郫县豆瓣、姜米、葱花炒香至油呈红色，入鲜汤、料酒、胡椒粉烧沸出味，沥去料渣，将汤汁倒入蒸碗，入乳鸽、银杏果，上笼蒸至软熟后捞出装盘；再将火腿片、冬笋片、水发香菇片放入锅中烧沸，入味精、芝麻油、水淀粉收汁至浓稠，浇淋在乳鸽上即成。

风味特色

色泽红亮，质地软糯，味道咸鲜微辣。

创意设计

银杏果在四川常于白露时节后采收。此菜以资阳产的乳鸽与银杏制作而成，当属典型的白露节气菜品。银杏果为药食两用之品，能敛肺定喘；乳鸽滋肾补气、解毒祛风，二者搭配，能益气补肾、养肺止喘。此外，银杏果还具有抗菌、抗过敏及降血压等功能；乳鸽所含支链氨基酸可促进机体蛋白质合成，加快伤口愈合。

Ingredients

2 pigeons; 30g glutinous rice; 50g ginkgo nuts; 15g ham, cut into cubes; 30g ham, sliced; 15g winter bamboo shoots, cut into cubes; 30g winter bamboo shoots, sliced; 15g water soaked shitake mushrooms, cut into cubes; 30g water soaked shitake mushroom, sliced; 500ml stock; 40g Pixian chili bean paste; 6g salt; 10ml Shaoxing cooking wine; 1g MSG; 0.5g pepper; 3ml sesame oil; 10g average batter; 10g ginger, finely chopped; 15g spring onions, finely chopped; 1,000ml cooking oil (70ml to be consumed)

Preparation

1. Boil the glutinous rice in a pot until almost cooked. Remove from the pot, add salt, ginkgo nuts, ham cubes, winter bamboo shoots cubes and mushroom cubes, and blend well to make the stuffing
2. Cut into the nape of the pigeons to de-bone, and rinse. Put the stuffing into the pigeon belly, and seal the cut with toothpicks. Blanch the pigeons and use the toothpicks to prick and deflate the pigeons.
3. Heat up the cooking oil to 120℃ in a wok, and add Pixian chili bean paste, ginger and spring onions. Stir them till aromatic and the oil turns to glistening red color. Add the stock, Shaoxing cooking wine and pepper, and bring to a boil. Strain the stock and discard the dregs. Pour the stock into a steaming bowl, add the

银杏乳鸽 Pigeon Stew with Ginkgo Nuts

pigeons and ginkgo nuts, and steam till the pigeon meat is soft and cooked through. Laddle out the pigeons, and place on a serving dish. Add the ham slices, winter bamboo shoots slices and mushroom slices to a wok, and bring to a boil. Add MSG, sesame oil and batter, and simmer till the sauce becomes thick. Pour the sauce over the pigeons.

Features

bright and lustrous colors; soft in texture; savory, scrumptious and spicy tastes

Notes

In Sichuan, ginkgo nuts are usually harvested after the White Dew. This typical White Dew dish is brown braised by using Ziyang pigeons and ginkgo nuts. Ginkgo nuts can be used as both food and medicine. They have the functions of nourishing the lung and calming asthma. Pigeon has the functions of nourishing the kidney, detoxifying the body and expelling rheumatism. Therefore, combining pigeon with ginkgo nuts can benefit qi, tonify the kidney, nourish the lung and relieve coughing. In addition, ginkgo nuts have anti-bacterial and anti-allergic features and can also prevent high blood pressure. Meanwhile, branched chain amino acid contained in pigeons helps to promote the synthesis of protein in human body and accelerate wound healing.

乐至麻饼

Lezhi Sesame Pies

食材配方

面粉100克　瓜子100克　花生碎100克　芝麻100克　葡萄干100克　冬瓜糖100克　橘红100克　白糖100克　食用油100毫升

制作工艺

① 将瓜子、花生碎、葡萄干、冬瓜糖、橘红拌匀制成馅心。

② 面粉加入食用油、白糖和成面团，下剂，按压成面皮，放入馅心包裹成团，再擀成圆饼形，撒上芝麻，入烤箱烘烤30分钟至色金黄、味酥香后取出装盘即成。

风味特色

色泽黄而不焦，皮酥馅脆，味道甜香。

创意设计

白露时节常与中秋相邻，乐至麻饼便是当地传统的中秋节著名食品。在养生方面，瓜子、花生健脾养胃，润肺化痰，芝麻润燥养阴，葡萄干养血滋阴，冬瓜糖、橘红化痰止咳，一同为馅则滋而不腻、润而不燥。此外，瓜子等食材富含不饱和脂肪酸，也有利于人体健康。

Ingredients

100g wheat flour; 100g sunflower seeds; 100g peanuts; 100g sesames; 100g raisins; 100g candied winter melon; 100g dried tangerine peels; 100g sugar; 100ml cooking oil

Preparation

1. Mix the sunflower seeds, peanuts, raisins, candied melon and dried tangerine peels. Stir well to make the stuffing.
2. Mix the wheat flour with cooking oil and sugar to make dough. Divide the dough into small portions, knead and press into wrappers. Wrap up some stuffing in a wrapper, flatten and roll with a rolling pin into round shapes. Drizzle with sesame seeds and bake for about 30 minutes till the pies are crispy, aromatic and golden. Transfer the pies to a serving dish.

Features

golden color; crispy texture; aromatic and sweet tastes

Notes

White Dew is usually close to the Mid-Autumn Festival. Lezhi Sesame Pie is one of the famous local traditional foods in the Mid-Autumn Festival. In food therapy, sunflower seeds and peanuts tonify the spleen, nourish the stomach, moisten the lung and reduce phlegm. Meanwhile, sesames help to generate body fluid and raisins help to nourish the blood. Candied melon and dried tangerine peels help to prevent phlegm from forming and stop coughing. In addition, sunflower seeds are rich in unsaturated fatty acid, which is beneficial to human health.

THE AUTUMN EQUINOX

秋分

流水今日 明月前身

点绛唇·金气秋分

宋·谢逸

金气秋分，风清露冷秋期半。
凉蟾光满，桂子飘香远。
素练宽衣，仙仗明飞观。
霓裳乱，银桥人散，吹彻昭华管。

秋分物候 | PHENOLOGY IN THE AUTUMN EQUINOX

秋分是二十四节气中的第十六个节气。当太阳运行到黄经180° 时即是秋分，时间为每年的9月22日~24日。据汉代董仲舒的《春秋繁露》记载："秋分者，阴阳相伴也，故昼夜均而寒暑平。"秋分之后，太阳直射地面的位置将继续南移，北半球昼短夜长的现象越来越显著，昼夜温差渐大，气温逐日下降，并意味着气候的变化开始步入深秋。

The the Autumn Equinox is the sixteenth of the 24 Solar Terms. It begins when the sun reaches the celestial longitude of 180°. It usually begins from 22nd to 24th September every year. Dong Zhongshu in Han Dynasty (202BC-220) has recorded in his book *Elaboration of Spring and Autumn Chronicles*, "the Autumn Equinox is accompanied by yin and yang; hence in the Autumn Equinox, day time and night time are equal, and coldness and hotness are equal too". After the Autumn Equinox, the position where the sun shines directly will move towards south. Night time will be longer and longer than day time in the Northern Hemisphere. The temperature difference between day and night will be getting large. The temperature is dropping day by day. It starts to enter the late autumn.

我国古代将秋分后的十五天分为三候：一候雷始收声；二候蛰虫坯户；三候水始涸。秋分节气之后，打雷的现象减少，虫类开始蛰伏在地下的洞穴中，而河流、湖泊的水量也开始明显减少。在四川，日平均气温将由此开始低于20℃。四川盆地内依然是秋雨绵绵。

Ancient China has divided the 15 days after the Autumn Equinox into 3 pentads: in the first pentad, the thunders begin to reduce; in the second pentad, insects begin to hide in the caves underground; in the third pentad, water in rivers and lakes has decreased significantly. In Sichuan, the daily average temperature will be below 20℃, and rain of autumn is ceaseless and lingering.

食材生产 | PRODUCING FOOD INGREDIENTS

秋分时节依然是农业种植最为忙碌的时期之一。东汉农学家崔寔在其《四民月令》中写道："凡种大小麦得白露节可中薄田，秋分中中田，后十日中美田。"秋分之后的四川地区，人们在收割秋天农作物的同时，还要随耕随种萝卜、莴笋、大头菜、青菜、白菜等多种蔬菜。在资阳地区，秋分时节有茄子、扁豆、南瓜、冬瓜、莲藕，以及雁江蜜柑、柠檬、通贤柚等多种食材上市，为节气菜点制作打下了良好的物质基础。

The Autumn Equinox is one the busiest periods for agriculture. Cui Shi, agronomist in the Eastern Han Dynasty (25-220), has written in his *Si Min Yue Ling*, "as for growing wheat and barley, farmers can sow them in barren fields in White Dew; farmers can sow them in normal fields in the Autumn Equinox; farmers can sow them in fertile fields 10 days after the Autumn Equinox". In Sichuan region, after the Autumn Equinox, people need sow vegetables like radish, lettuce, kohlrabi, pakchoi, Chinese cabbage, in addition to harvesting autumn crops. In Ziyang region, eggplant, lentil, pumpkin, winter melon, lotus root, Yanjiang tangerine, lemon, Tongxian pomelo are on the market in the Autumn Equinox, which have made good preparations for making seasonal dishes.

饮食养生 | DIETARY REGIMEN

秋分过后，我国大部分地区日照减少，气温降低，人们在养生方面要注意保持身体的阴阳平衡和饮食多样性。秋分之后，四川多地气温降低、夹杂雨水，往往会出现湿冷状况，更适合采用《黄帝内经》中所说 “秋多辛” 的调和原则。四川自古就有“尚滋味，好辛香”的饮食传统，秋分时节可选用花椒、辣椒等辛温之品温以祛寒，辛以发散，有利于祛除湿冷。但是，辛温之品的使用必须适当，不能过度，以防阳气过于耗散，这便是唐代孙思邈在《千金方》中提出的秋季“省辛增酸”之意，因而川菜味型中的酸辣味型就尤其适合。当雨水减少，气候逐渐趋于干燥时，饮食上就应注重养护阴液。萝卜、莲藕、梨、苹果、银耳等均能养阴润燥。若是选用单纯带辣味的菜品时，应搭配酸味、酸甜味型的菜品等，如酸萝卜老鸭汤、银耳百合羹，以及山药、时令水果入菜，或用豆浆入羹入点。此外，由于秋天人们的肠胃功能较弱，因此可多食用茯苓、芡实、山药等健脾益肾，培补后天脾胃，使水谷精微充盛，为冬季进补打下基础。

After the Autumn Equinox, the sunshine has decreased and temperature has dropped in many parts of China. People should pay attention to keep human body's balance between yin and yang, and the varieties of diets. In Sichuan, there is a famous poem by Li Shangyin in Tang Dynasty, "Autumn rain comes at night and the pool gets full." The dropping temperature with rain have caused clamminess after the Autumn Equinox. Hence, people should obey the principle of eating more pungent foods in autumn from *Inner Canon of Huangdi*. Sichuan has the eating customs of preferring to good taste and spicy. Taking pungent foods like Sichuan pepper and chilly can dispel coldness and dampness in the Autumn Equinox. However, people should eat proper amount of pungent and warm food, and should not eat too much to avoid waste too much positive energy. Sun Simiao in Tang Dynasty (618-907) has said in his *Thousand Golden Prescriptions* that people should eat less pungent food and more sour food in autumn, so the sour and hot dishes from Sichuan cuisine are particularly appropriate. When rainfall decreases and climate is getting dry, the diet should pay attention to maintain body liquor with the nature of yin. Radish, lotus root, pearl, apple, tremella can tonify the yin and moisten the dryness. When people choose pure pungent dishes, it is better to add food ingredients with the tastes of sour or sour and sweet, like Old Duck Soup with Sour Radish, Tremella and Lily Soup, or to add some Chinese yam, seasonal fruits, or add some soybean milk into thick soup and snacks. Moreover, human gastrointestinal function is a little weak in autumn. Therefore, they should eat more tuckahoe, gorgon fruit and Chinese yam to strengthen spleen and stomach, to tonify kidney, and lay a foundation for further food supplement in winter.

秋分美食 | FINE FOODS IN THE AUTUMN EQUINOX

俗语道，“秋风起，蟹黄肥”，曾有晋朝名人毕卓“一手持蟹螯，一手持酒杯，拍浮酒池中,便了足一生”成为美谈。秋分时节，在四川地区，除了各种农作物迎来丰收外，同时也是螃蟹最肥美、最滋补的时候。近年来，吃螃蟹在四川民间十分流行，许多人将螃蟹清蒸或炒制，成为秋分时节前后的重要美食。这一时段，马齿苋等一些野菜也登堂入室，成为四川百姓餐桌上的常客，吃法多以凉拌为主。

As the proverb goes, "crab is fat when autumn wind begins". Bi Zhuo, an eminent person in Jin (266-420) said "Lifetime is fully satisfying when you have enough wine to drink with one pincer in one hand, and one wine cup in another hand." In the Autumn Equinox, in addition to good harvest of all kinds of crops, the crab is most delicious and nutritious. In recent years, eating crab is very popular in Sichuan. Steamed crab and fried crab are important dishes around the Autumn Equinox. In this period, potherbs like purslane, which is eaten cold with sauce, has become normal dishes for Sichuan people.

青花椒焖蟹

Stewed Crabs with Green Sichuan Peppercorns

食材配方

螃蟹200克　年糕60克　水发雪魔芋60克　鲜青花椒30克　小青椒20克　小米辣15克　野山椒15克　大蒜15克　鲜汤100毫升　蚝油10毫升　酱油5毫升　鸡精1克　味精1克　白糖2克　青花椒油3毫升　食用油1000毫升（约耗80毫升）

制作工艺

① 螃蟹经初加工后洗净，斩成块；年糕、水发雪魔芋切成菱形块；小青椒、小米辣切成节。

② 锅置火上，入油烧至150℃，放入螃蟹炸至色红后捞出。

③ 锅置火上，入油烧至150℃，放入大蒜、鲜青花椒、野山椒、小青椒、小米辣、螃蟹炒香，入鲜汤、年糕、雪魔芋、食盐烧沸，改用小火焖10分钟，入蚝油、鸡精、味精、白糖、酱油收汁，入青花椒油炒匀装盘即成。

风味特色

色彩鲜艳，质感丰富，味道咸鲜麻香略辣。

创意设计

秋分时节菊黄蟹肥，二者相得益彰。此时，以螃蟹为主料，与年糕、雪魔芋焖制，用乐至青花椒、辣椒调味，味道咸鲜，麻香微辣。在养生方面，蟹肉性寒、清热解毒，调以性热的青花椒及辣椒，一是体现了川菜的风味特色，二能平调阴阳。此外，蟹肉是高蛋白、低脂肪食材，且富含钙质等，有预防心脑血管疾病的功效。

Ingredients

200g crabs; 60g new year cake; 60g water soaked konjak jelly; 30g green Sichuan peppercorns; 20g small green chilies; 15g bird's eye chilies; 15g mountain chilies; 15g garlic; 100ml stock; 10ml oyster sauce; 5ml soy sauce; 1g chicken essence; 1g MSG; 2g sugar; 3ml green Sichuan pepper oil; 1,000ml cooking oil (80ml to be consumed)

Preparation

1. Pre-process the crabs, rinse and chop into chunks. Cut the rice cake and konjak jelly into diamond chunks. Cut the green chilies and bird’s eye chilies into small segments.
2. Heat up the cooking oil to 150℃ in a wok, and deep fry the crabs till red.
3. Heat up the cooking oil to 150℃ in a wok, add garlic, green Sichuan peppercorns, mountain chilies, green chilies, bird's eye chilies and crabs, and stir fry till aromatic. Blend in the stock, rice cake, konjak jelly and salt, and bring to a boil. Cover, simmer over a low flame for about ten minutes, and add the oyster sauce, chicken essence, MSG, sugar and soy sauce. Continue to simmer to reduce the sauce. Add the green Sichuan pepper oil, and blend well. Transfer to a serving dish.

Features

bright color; salty, delicious, numbing and spicy tastes

Notes

In the Autumn Equinox, the chrysanthemums are in bloom while the crab meat is delicious. In this dish, crab is stewed with new year rice cake and konjak jelly, and seasoned by the green Sichuan pepper (produced in Lezhi) and mountain chilies. In food therapy, crab meat is cold in nature and it has a function of clearing away inner heat and detoxify the body. The green Sichuan pepper and chilies, which are representative of Sichuan, are hot in nature and can balance the coldness (yin) of the crab. In addition, crab meat has a function of preventing cerebrovascular diseases, for it is rich in protein and calcium and low in fat.

食材配方

老南瓜400克　咸蛋黄50克　南瓜泥30克　白糖2克　味精3.5克　肉松20克　黄油20克　低筋面粉100克　食用油1000毫升（约耗50毫升）

制作工艺

① 老南瓜改刀成约4厘米见方的块，汆水后捞出沥干水分，裹上面粉。

② 锅置火上，放入食用油烧至150℃，放入南瓜炸至定型、表皮酥脆时起锅。

③ 锅置火上，放入黄油，下咸蛋黄、南瓜泥、白糖炒香，加入南瓜炒匀，起锅装盘，撒上肉松即成。

风味特色

色泽金黄，质地香糯，味道咸甜。

创意设计

秋分时节，在四川地区，人们除了食用辛香食物以祛湿外，还应适当食用富含膳食纤维的时令瓜果、蔬菜。金沙南瓜咸甜香糯，可为常食之品。在养生方面，南瓜健脾益气、解毒消肿，蛋黄养阴润燥、补肾益精，二者功效互助、咸甜互补。此外，南瓜富含纤维素，其南瓜多糖能提高人体免疫机能。

金沙南瓜

Golden Pumpkin

Ingredients

400g pumpkin; 50g salted egg yolk; 30g smashed pumpkin; 2g sugar; 3.5g MSG; 20g pork floss; 20g butter; 100g cake flour; 1,000ml cooking oil (about 50ml to be consumed)

Preparation

1. Cut the pumpkin into 4 cm cubes, blanch and drain. Coat the cubes with cake flour.
2. Heat oil in a wok to 150℃, and deep-fry the pumpkin cubes until crispy.
3. Place a wok over the flame, add the butter, salted egg yolk, smashed pumpkin and sugar, and stir fry till fragrant. Blend in the pumpkin cubes, and stir well. Transfer to a serving dish and sprinkle with the pork floss.

Features

golden in color; glutinous in texture; salty and sweet in taste

Notes

On the Autumn Equinox, Sichuan people eat not only spicy food to eliminate body dampness but also seasonal melons, fruits and vegetables that are rich in dietary fibers. Golden Pumpkin tastes salty, sweet and glutinous and is favored by people in Sichuan. The dish is also great for people's health. Pumpkin tonifies the spleen and replenishes qi, helps detoxification and reduces body swelling. Salted egg yolk nourishes yin, moistens the dryness, and reinforces the kidney. These two ingredients are complementary in medical functions and tastes. Besides, pumpkin contains rich cellulose as well as polysaccharide which strengthens people's immunity function.

荠菜烧麦

Shepherd's Purse Shaomai

食材配方

面粉200克　猪肥肉100克　猪瘦绞肉150克　荠菜200克　细淀粉100克　食盐5克　胡椒粉1克　味精2克　酱油6毫升　芝麻油5毫升

制作工艺

① 面粉加入冷水调制成硬度合适的软面团，盖上湿毛巾醒面。

② 猪肥肉煮熟，晾冷后切成小颗粒；荠菜洗净，入沸水中焯水，取出漂冷后切成细末，挤去水分；猪瘦绞肉中加味精、酱油、芝麻油、胡椒粉、食盐拌和均匀，再下熟肥肉粒和荠菜末拌匀制成馅料。

③ 将醒好的面团搓成圆条、下剂，压扁后擀成圆皮，粘上细淀粉，6～8张叠码整齐，置案板上捶打面皮的边沿，使其形成中间厚、边沿薄的荷叶形圆皮。

④ 取面皮装上馅心，捏成白菜形成烧麦生坯，入笼中用旺火蒸3分钟左右揭开笼盖，洒上少许冷水，再盖上笼盖蒸熟即成。

风味特色

造型美观，味道咸鲜，荠菜香浓。

创意设计

荠菜是四川一些地方春秋两季皆有出产的野菜之一，将其与猪肉一起做成馅料，制成烧麦是野菜入点的创新。成菜荤素搭配，营养合理，咸鲜香浓。在养生方面，荠菜凉肝、清热，猪肉补肾滋阴、益气养血，二者搭配制成馅料，经面皮包裹能最大程度地保存其维生素，同时能养肝养血，益于健康。

Ingredients

200g wheat flour; 100g pork fat; 150g minced lean pork; 200g shepherd's purse; 100g cornstarch; 5g salt; 1g pepper; 2g MSG; 6ml soy sauce; 5ml sesame oil

Preparation

1. Mix the flour and water to make soft dough. Cover it with a wet towel and let stand in room temperature.
2. Boil the pork fat thoroughly and ladle out. Leave to cool and mince. Rinse the shepherd's purse, and blanch. Transfer to cold water, and leave to cool. Mince and squeeze off water. Mix the minced lean pork with MSG, soy sauce, sesame oil, pepper and salt, add the minced pork fat and shepherd's purse, and mix well to make the stuffing.

3. Knead the dough into long strips, cut into small lumps, press and roll out into round wrappers. Sprinkle with cornstarch, and lay every 6 to 8 wrappers in a stack on a cutting board. Pound the edges of the stack so that the wrappers look like lotus leaves that are thick in the middle but thin on the edges.
4. Fold the stuffing up with wrappers, and shape into Chinese cabbage. Steam for about 3 minutes over high heat, remove the cover, and sprinkle with some cold water. Cover, and continue to steam till cooked through.

Features

beautiful design; salty and fresh tastes with a strong aroma from the shepherd's purse

Notes

Shepherd's purse is one of the wild vegetables growing in spring and autumn. What makes the dish special is the stuffing made from shepherd's purse and pork. With meat and vegetables, this dish has balanced nutrition and tastes salty and fresh. It is also great for people’s heath. Shepherd's purse clears heat and relives the liver. Pork nourishes the kidney and blood, and reinforces qi. Both ingredients are wrapped up in dough to prevent vitamin loss to the largest extent.

咏廿四气诗·寒露九月节
唐·元稹
寒露惊秋晚，朝看菊渐黄。
千家风扫叶，万里雁随阳。
化蛤悲群鸟，收田畏早霜。
因知松柏志，冬夏色苍苍。
COLD DEW

寒露

秋菊染黄

风清露冷

寒露物候 | PHENOLOGY IN COLD DEW

寒露是二十四节气中的第十七个节气。当太阳运行到黄经195° 时为寒露，时间为每年的10月8日~9日。“寒”指是指寒冷，“露”是露水，古代通常用“露”来表达天气转凉变冷之意。元朝吴澄《月令七十二候集解》言：“寒露，九月节。露气寒冷，将凝结也。”意思是说到了寒露，天气变冷，地面上的露水快要凝结成霜了。

Cold Dew is the seventeenth of the 24 Solar Terms. It begins when the sun reaches the celestial longitude of 195°. It usually begins from 8th to 9th October every year. Chinese word “Han” means cold and “Lu” means dew. In ancient times, people use dew to express the meaning of turning cold. In his book *A Collective Interpretation of the Seventy-Two Phenological Terms*, Wu Cheng (Yuan Dynasty 1271-1368) said, “Cold Dew is the September (lunar calendar) Festival. In Cold Dew, it is getting colder and the dew on ground will turn into frost.”

我国古代将寒露后的十五天分为三候：一候鸿雁来宾；二候雀入大水为蛤；三候菊有黄华。寒露到来，北方的气候开始变得寒冷，候鸟已迁徙到南方；雀鸟也销声匿迹，海边却出现了与雀鸟颜色条纹相似的蛤蜊，民间传说这是雀鸟变的；菊花也迎来了花期。在四川盆地，这时日平均气温大多降为15℃～17℃。

Ancient China has divided the 15 days after Cold Dew into 3 pentads. In the first pentad, the north is getting cold and migrant birds have flown to south. In the second pentad, birds have hidden, and clam with the markings of birds have appeared in the shore, which are believed to be changed from birds. In the third pentad, chrysanthemum blossoms. In Sichuan basin, the daily average temperature is 15～17℃。

食材生产 | PRODUCING FOOD INGREDIENTS

随着寒冷天气的到来，包括四川在内的广大西南地区，除了继续进行秋收外，还要抓紧时间适时播种莲花白、油菜、莴笋、菠菜、辣椒、蚕豆、豌豆等蔬菜作物。在资阳，莲藕、南瓜、冬瓜、茄子等蔬菜，以及雁江蜜柑、柠檬、通贤柚等深受广大消费者欢迎的特色食材，依然处于继续收获中。

With the arrival of cold weather, people have to seize the time to sow cabbage, winter rape, lettuce, spinach, chilly, broad bean, pea, in addition to autumn harvest in vast Southwest China including Sichuan. In Ziyang, people still harvest lotus root, pumpkin, winter melon, eggplant, and Yanjiang tangerine, lemon, Tonxian pomelo which are very popular among customers.

饮食养生 | DIETARY REGIMEN

寒露时节，气温比白露时更低，四川盆地阴雨天气依然较多。在饮食养生方面，首先要确保食材干净、无污染，还要去除阴雨天可能的霉变之物。其次，日常饮食要以清淡、易消化的食物为主，宜选择高蛋白、脂肪适量、富含维生素的食物，忌暴吃冷饮。烹调上应少煎炸，控制咸味调味品的使用量。再次，应选择食用一些健脾开胃、润肺养阴的食物，如土豆、南瓜、莲子、梨、柑橘、葡萄、柿子等。

Temperature in Cold Dew is lower than in White Dew. It is raining frequently in Sichuan basin. As for current diet, people should keep food ingredients clear and free of contamination first, and get rid of foods with mould caused by wet weather. Then, daily diet should be mainly composed of light and digestible foods. People should choose foods with high protein, less fat and abundant vitamins; and should not take too much cold drink. As for the way of cooking, we should less fry and control the use of salty spices. Furthermore, people should take some foods which can tonify spleen and stomach, moisten lung and nourish our energy, like potato, pumpkin, lotus seed, pearl, orange, grape, persimmon, etc.

寒露美食 | FINE FOODS IN COLD DEW

寒露节气的时间，大多在农历的九月。古人有诗言，“九月团脐十月尖，持螯饮酒菊花天”，螃蟹和菊花酒搭配食用，是这个时节最常见的饮食习俗。同时，寒露节气的时间在许多年份与农历九月初九的重阳节相近，除了饮菊花酒，还有菊花糕等美食。据清朝道光年间的《新都县志》载：“重阳，登高啖花糕，酌菊酒。”“花糕”，又称五色糕、重阳糕，是用糙米、红枣、核桃、果脯等食材制成。到了民国时期，菊花酒的酿造开始逐渐减少，但却出现了一种用菊花与蔬菜、肉类等制成的“菊花锅”。民国时期的《新繁县志》就记载说：“菊花酒者，采菊茎叶，杂黍米酿之。今造此酒者甚少，间有以菊花芼羹者，曰‘菊花锅’。”时至今日，到了寒露时节，市场上丰富的食材可为民众的饮食提供更多的选择。

Cold Dew is mostly in September of lunar calender. There was old poem said, “September and October crab is nice, holding pincers drinking chrysanthemum wine is just perfect”. Eating crabs with chrysanthemum wine is the most popular custom. Cold Dew is close to Double Ninth Festival in many years. In addition to chrysanthemum wine, there are fine foods like Chrysanthemum Cake. According to the *Xindu Local Chronicles* in Daoguang Period Qing Dynasty(1636-1912), people would climb mountains, eat flower cake and drink chrysanthemum wine. Flower cake, which is called five colors cake or double ninth cake, is made of brown rice, red date, walnut, preserved fruits. In the period of the Republic of China, people brewed less chrysanthemum wine, but began to make a kind of food which made of chrysanthemum, vegetables and meat. *Xinfan Locl Chronicles* in the period of the Republic of China recorded, “People brew chrysanthemum wine with chrysanthemum stem and leaves, millet. Currently, people brew very less chrysanthemum wine, but make chrysanthemum pot.” Nowadays, abundant food ingredients on the market have provided more options for common people.

菊花鱼 | Chrysanthemum Fish

食材配方

草鱼1尾（约2000克） 食用鲜菊花瓣20克 黄瓜200克 食盐10克
料酒15毫升 姜片10克 葱段15克 浓缩果汁200毫升 白糖200克
清水100毫升 干淀粉200克 食用油1500毫升（约耗100毫升）

制作工艺

① 草鱼去头、尾和鱼骨，取鱼肉剞成“十”字花刀，再改刀成5厘米长的段，加入食盐、料酒、姜片、葱段拌匀，码味15分钟。

② 锅置火上，入油烧至160℃，将鱼肉放在干淀粉中裹匀，入锅炸至定型、成熟后捞出；待油温回升至200℃时，将鱼肉入锅复炸至外酥时捞出，放入垫有黄瓜片的盘中。

③ 锅置火上，入清水、浓缩果汁、白糖收汁浓稠，浇淋在鱼花上，撒上鲜菊花瓣即成。

风味特色

形似菊花，色泽橙黄，外酥内嫩，味道酸甜，果香浓郁。

创意设计

寒露时节，天气持续转凉，菊花依然绽放，沱江里的鱼愈加肥美。将鱼肉做成菊花形，配上金黄色的汁与食用菊花，色泽鲜艳，清爽温暖。在养生方面，草鱼利水消肿、健脾祛湿；菊花疏风散热、清热解毒，能缓解炸制鱼肉的燥热。此外，草鱼具有高蛋白、低脂肪的特点；菊花有抑菌、抗炎、降低血压和调节免疫力之功。

Ingredients

1 grass carp (about 2,000g); 20g edible chrysanthemum; 200g cucumbers; 10g salt; 15ml Shaoxing cooking wine; 10g ginger, sliced; 15g spring onions, segmented; 200ml condensed orange juice; 200g sugar; 100ml water; 200g cornstarch; 1,500ml cooking oil (about 100ml to be consumed)

Preparation

1. Remove the head, tail and bones of the fish to get fillets. Cut crosses on the fillets and chop into 5cm long segments. Add the salt, cooking wine, ginger and spring onions, mix well, and leave to marinate for 15 minutes.
2. Heat cooking oil in a wok to 160℃. Roll the fish in the cornstarch, deep fry until stiff enough and ladle out. Heat oil to 200℃, deep fry the fish until crispy and transfer to a serving plate with sliced cucumbers laid on the bottom.
3. Heat water, orange juice, and sugar in a wok for reduced sauce. Drizzle the sauce over the fish and sprinkle with the chrysanthemum.

Features

pleasant chrysanthemum-like form; gold in color; sour, sweet, fresh tastes with strong fruit fragrance; crispy outside and tender inside.

Notes

When Cold Dew comes, it becomes cold. In this season, chrysanthemum is still in blossom and Tuojiang grass fish grows to be a great cooking ingredient. Deep-fried fish, drizzle with reduced source and garnished with edible chrysanthemum, looks golden and tastes fresh. The dish is great for people's health. Grass fish promotes water excretion, relieves edema, tonifies the spleen and clears body dampness. Golden chrysanthemum helps clear body heat and reduces the greasiness of fried fish. In addition, Tuojiang fish has rich protein and less fat. Chrysanthemum inhibits the bacteria, diminishes inflammation, reduces blood pressure and improves immunity.

怪味果仁

Strange Flavor Nuts

食材配方

核桃仁100克　花生仁100克　清水200毫升　白糖200克　甜面酱10克　食盐2克　辣椒粉5克　花椒粉1克　柠檬酸0.1克　食用油500毫升（约耗20毫升）

制作工艺

① 核桃仁用沸水浸泡后撕去外皮，入锅炸酥后捞出。

② 花生仁入烤箱烤至酥香，取出晾冷后去皮。

③ 锅置火上，入清水、白糖，用中小火慢慢熬至糖液黏稠，再入甜面酱搅匀，至水分将干时，放入食盐、辣椒粉、花椒粉、柠檬酸搅匀，端离火源，待糖液中的泡沫消失后放入核桃仁、花生仁炒匀，起锅，待其晾凉成霜状后装盘即成。

风味特色

质地酥脆，味道甜香麻辣。

创意设计

寒露时节，各种坚果开始陆续上市，以核桃、花生、板栗等最为常见，采用“糖粘”这一川菜特色烹调方法制成的怪味果仁，是佐酒佳品。在养生方面，花生补肺化痰，核桃补肾益精、温肺定喘，二者富含不饱和脂肪酸，尤其是核桃还含有少见的α－亚麻酸，有增强血管弹性、抗氧化、降血压及减轻脂肪肝之功。

Ingredients

100g walnut kernels; 100g peanuts; 200ml water; 200ml sugar; 10g fermented flour paste; 2g salt; 5g ground chilies; 1g ground Sichuan pepper; 0.1g citric acid; 500ml cooking oil (about 20ml to be consumed)

Preparation

1. Soak the walnut kernels in boiling water and remove peels. Deep fry the kernels till crispy, and remove from the oil.
2. Put the peanuts in the oven and roast till crispy. Remove, leave to cool and peel.
3. Heat water and sugar in a wok over a medium-low flame until the sugar melts and the juice is sticky and thick. Add the fermented flour paste, and stir to evaporate the moisture. Add the salt, ground chilies, ground Sichuan pepper and citric acid, stir well and remove from the flame. Wait till the bubbles disappear, and blend in the walnut kernels and peanuts. Stir till the sugar frosts, leave to cool and transfer to a serving dish.

Features

crispy texture; sweet and spicy tastes

Notes

When Cold Dew comes, various nuts appear on the table. Among them, walnuts, peanuts and chestnuts are the most popular. Cooked in a Sichuan way, the nuts are preferred as snacks when people drink beer or spirits. This dish is also great for human health. Peanuts reinforce the lung and reduce phlegm; walnuts reinforce the kidney and lung, and relieve asthma; and both of them have rich unsaturated fatty acid. Besides, walnuts have the rare α-linolenic acid which increases vascular elasticity, prevents oxidation, and reduces blood pressure and liver fat.

重阳糕 Double Ninth Cake

食材配方

低筋面粉360克　泡打粉10克　细砂糖110克　牛奶300毫升　鸡蛋清60毫升　食用油50毫升　酵母1.5克　干桂花2克　葡萄干30克

制作工艺

①将牛奶、鸡蛋清、细砂糖混合搅拌至砂糖完全化开，先加入低筋面粉、泡打粉、酵母混合搅拌均匀，再加入食用油、干桂花搅拌均匀。

②模具上刷油，将面糊倒入其中，高度大约2厘米，放入醒发箱醒发30分钟后取出，在表面撒上葡萄干，用大火蒸20分钟至成熟，放凉后切成方块装盘即成。

风味特色

色泽微黄，质地松软，味道甘甜，桂花香浓。

创意设计

寒露常与重阳节比邻。重阳糕是重阳节的传统节令食品，通常以面粉为主料制成。此糕在常见做法的基础上进行创新，在表面撒上干桂花，色彩更丰富，质地更松软，味道更甜香。在养生方面，桂花温肺散寒，面粉益气和中，经发酵后制成重阳糕，会产生大量B族维生素，不仅是应季美点，也更具食疗保健价值。

Ingredients

360g cake flour; 10g baking powder; 110g caster sugar; 300ml milk; 60ml egg white; 50ml cooking oil; 1.5g yeast; 2g dried osmanthus flowers; 30g raisin

Preparation

1. Mix the milk, egg white with sugar. Add the cake flour, baking powder and yeast and mix well. Add cooking oil and dried osmanthus flowers and mix well to make a paste.
2. Brush the mould with oil, and pour the paste 2cm deep into the mould. Put it in the proofer and let stand for 30 minutes. Remove from the proofer, sprinkle with raisins and steam over a high flame for 20 minutes till cooked through. Remove from the steamer, leave to cool, cut into squares and transfer to a serving dish.

Features

light yellow in color; soft and sweet taste; a scent of osmanthus flower

Notes

When Cold Dew comes, Double Ninth Festival is not far behind. Double Ninth Cake is especially made on this occasion. With dried osmanthus flowers sprinkled on its surface, the soft cake tastes fresh and sweet. Double Ninth Cake also has great heath benefits. Osmanthus flowers warm the lung and dispel inner cold. Wheat flour strengthens qi, and becomes rich in vitamin B after being fermented. Double Ninth Cake is not only a seasonal dessert, but also a health food.

FROST'S DESCENT

都门霜降日作

明·郑茂

风雨连朝动客愁，
笳声呜咽满边楼。
卷帘何事看新月，
一夜霜寒木叶秋。

霜降

林寒涧肃　一叶知秋

霜降物候 | PHENOLOGY IN FROST'S DESCENT

霜降是二十四节气中的第十八个节气。当太阳运行到黄经210° 时为霜降，时间在每年的10月23日～24日。元朝吴澄《月令七十二候集解》言："霜降，九月中。气肃而凝露结为霜矣。"到农历九月，天气变冷，空气中的水分已经凝结成霜。

Frost's Descent is the eighteenth of the 24 Solar Terms. It begins when the sun reaches the celestial longitude of 210°. It usually begins from 23rd to 24th October every year. In his book *A Collective Interpretation of the Seventy-Two Phenological Terms*, Wu Cheng (Yuan Dynasty 1271-1368) said that in the ninth lunar month, the weather was getting cold and the molecule of water in air has turned into frost.

我国古代将霜降后的十五天分为三候：一候豺乃祭兽；二候草木黄落；三候蛰虫咸俯。随着天气转凉，豺狼开始大量捕获猎物，囤积脂肪为过冬做准备；草木的枝叶开始衰败枯萎；虫类则开始蛰伏地下，等待春天的到来。在四川，尚无冰霜出现，但雨量减少，大致仅有15～50毫米。

Ancient China has divided the 15 days after Frost's Descent into 3 pentads: in the first pentad, wolves have hunted more preys and stored fat for winter; in the second pentad, plants begin to wither; in the third pentad, worms and insects start to hide underground waiting for next spring. In Sichuan, there is no frost then, but the rainfall have been decreased to 15mm to 50mm.

食材生产 | PRODUCING FOOD INGREDIENTS

在霜降时节，四川地区依然处于农作物的收获之际。据清朝嘉庆年间的《三台县志》载："'霜降节'而无霜、风，以地气燠耳。江水渐涸，木叶初凋，百谷登场，百果实。"清朝道光年间的《隆昌县志》则具体写到："收皂角、木瓜，种蚕豆、蒜、芥。"在资阳，人们不仅能收获四季豆、南瓜、冬瓜、柠檬等蔬果，还有鸽子、泥鳅、鹌鹑等食材可用于制作菜点，同时还要种植萝卜、莴笋、莲花白、白菜等蔬菜，为冬天也能食用到较多的蔬菜做好准备。

In Frost's Descent, it is still harvest time for Sichuan region. *Santai County Chronicles* in Jiaqing Period Qing Dynasty recorded, "There is no frost and wind in Frost's Descent because the land is still warm. The water of river is decreasing, plants begins to wither, all the grains and all kinds of fruits are in harvest time." *Longchang County Chronicles* in Daoguang Period Qing Dynasty recorded, "people harvest saponin, papaya, and plant broad bean, garlic, mustard. " In Ziyang, pepole not only harvest kidney bean, pumpkin, winter melon, lemon, but also make dishes with pigeon, loach, quail. Meanwhile, they will plant radish, lettuce, cabbage, Chinese cabbage for the coming winter.

饮食养生 | DIETARY REGIMEN

霜降正是季节变换的时节，天气渐冷，气温变化剧烈、昼夜温差增大，但是在南方地区尚未到初霜时期。民谚说“一年补透透，不如补霜降”，霜降进补时机正好，但应平补，适当养肾健脾，为冬季进一步进补做好准备。在饮食养生上，应少食寒凉食物，由于天气相对干燥，可食用梨、苹果、柿子、百合、莲藕、荸荠等生津润燥的食物，也可多吃板栗补脾益肾。此外，吃红薯、山药、芋头等能养阴益气，吃白果能敛肺止泻，可以制作板栗烧鸡、白果炖鸡、陈皮萝卜炖牛肉、山药牛肉，还可以用芋头、红薯、莲藕等入菜做点。

Frost’s Descent is just at the turn of seasons. It is getting cold; temperature changes dramatically; temperature difference between day and night is increasing. However, there is not frost in the South. As the proverb goes, “Frost’s Descent is the right time to get tonic foods.” People should take proper nourishment to tonify kidney and spleen, and make preparations to further tonify human body in winter. As for the diet, people should eat less cold food, and take more pearl, apple, persimmon, lily, lotus root, which can engender liquor in dry weather, and eat Chinese chestnut to tonify spleen and kidney. Moreover, sweet potato, Chinese yam, taro can nourish our energy; ginkgo can constrain lung and stop diarrhea. People can make Braised Chicken with Chinese Chestnut, Stewed Chicken with Ginkgo, Beef Stewed with Tangerine Peel and Radish, Beef with Chinese Yam. Taro, sweet potato and lotus root can be added into dishes as well.

霜降美食 | FINE FOODS IN FROST'S DESCENT

霜降时节，四川民间也有一些相应的饮食习俗。与资阳邻近的隆昌地区，人们就曾于霜降日制作茱萸油。据清朝道光年间的《隆昌县志》载：“霜降日，农人涤耒耜藏于室。制萸，谓之‘艾油’，供养老人，备御寒具。”此外，霜降时节，四川民间还有吃兔肉的习俗，正如民谚所言：“迎霜兔肉来秋补。”此时的兔肉味道鲜美、营养价值高，还有滋补防病、延年益寿的功效。

In Frost’s Descent, there are related diet customs. In Longchang region next to Ziyang, people once made cornel oil on the day of Frost’s Descent. *Longchang County Local Chronicles* in Daoguang Period Qing Dynasty recorded, “On the day of Frost's Descent, farmers clean tilling tools and store them indoors. People make cornel oil for the elderly and prepare appliance for winter.” Furthermore, there is a custom of eating rabbit in Frost’s Descent. As the proverb goes, “Intake of rabbit meat benefits the body during autumn Frost’s Descent”. Rabbit meat in this period is delicious and nutritious, which can prevent disease and prolong life.

鲜椒嫩兔

Tender Rabbit with Chili Peppers

食材配方

兔肉400克　仔姜100克　青尖椒100克　红尖椒100克　鲜青花椒20克
郫县豆瓣30克　泡辣椒末30克　食盐3克　料酒30毫升　水淀粉20克　味精2克
芝麻油3毫升　青花椒油5克　鲜汤500毫升　食用油1000毫升（约耗150毫升）

制作工艺

① 兔肉切成约1.2厘米见方的小丁，加入食盐、料酒、水淀粉拌匀。
② 仔姜切成约1厘米见方的丁；青尖椒、红尖椒分别切成长约1厘米的节。
③ 锅置火上，入油烧至120℃时，放入兔丁滑熟后捞出。
④ 锅置火上，入油烧至120℃时，放入郫县豆瓣、泡辣椒末、鲜青花椒炒香，再下仔姜、青红尖椒、兔丁炒匀，掺入鲜汤烧沸，最后下料酒、味精、芝麻油、青花椒油烧至入味起锅装盘即成。

风味特色

色泽红亮，味道鲜辣突出，兔肉细嫩。

创意设计

四川民间在霜降时节有吃兔肉的习俗，正如民谚所言，“迎霜兔肉来秋补”。在养生方面，兔肉味甘、性寒，辅以辛温的辣椒、花椒，寒热中和，既体现了川菜的特有风味，也利于此时的平补。同时，兔肉是典型的高蛋白、低脂肪食物，有美容养颜等作用。

Ingredients

400g rabbit; 100g tender ginger; 100g green chili peppers; 100g red chili peppers; 20g green Sichuan peppercorns; 30g Pixian chili bean paste; 30g pickled chilies, finely chopped; 3g salt; 30ml Shaoxing cooking wine; 20g average batter; 2g MSG; 3ml sesame oil; 5ml green Sichuan pepper oil; 500ml stock; 1,000ml cooking oil (about 150ml to be consumed)

Preparation

1. Cut the rabbit into 1.2cm cubes, and mix well with the salt, cooking wine and batter.
2. Cut the ginger into 1cm cubes and the green and red chili peppers into 1cm long sections.
3. Heat oil in a wok to 120℃, stir fry the rabbit cubes till cooked through and remove.
4. Heat oil in a wok to 120℃, add the chili bean paste, pickled chilies, green Sichuan peppercorns, ginger, red chili peppers, green chili peppers and rabbit, stir well and add the stock. Bring to a boil, and add cooking wine, MSG, sesame oil and green Sichuan pepper oil. Continue to braise till the rabbit has absorbed the flavors of the ingredients, and transfer to a serving dish.

Features

reddish brown color; savory and spicy tastes; tender rabbit meat.

Innovation

"Intake of rabbit meat benefits the body during autumn Frost's Descent." The saying suggests the folk custom of having rabbit during the solar term of Frost's Descent'. In terms of food therapy, rabbit has a cooling function while chilies and Sichuan pepper are warm in nature, so their combination achieves a balanced effect and promotes body function. Besides, rabbit is high in protein and low in fat, a perfect food for beauty and skin.

鱼香藕夹

Fish Flavor Lotus Root Pie

食材配方

鲜藕200克　猪肉80克　鸡蛋液140毫升　淀粉100克　泡辣椒末40克　姜米5克　蒜米15克　葱花25克　食盐3克　料酒10毫升　白糖20克　醋10毫升　酱油5毫升　味精1克　水淀粉15克　鲜汤100毫升　食用油1500毫升（约耗80毫升）

制作工艺

① 鲜藕去皮，切成连夹片；用100毫升鸡蛋液与淀粉调匀成全蛋淀粉糊。

② 猪肉剁碎，加入食盐、料酒和余下的鸡蛋液拌匀调成肉馅，酿入藕片中，制成藕饼初坯。

③ 食盐、白糖、料酒、醋、酱油、味精、水淀粉、鲜汤入碗调成芡汁。

④ 锅置火上，入油烧至150℃时，将藕饼初坯裹上全蛋淀粉糊，入锅炸至定型后捞出；待油温回升至200℃时，再将藕饼初坯入锅中炸至外酥内熟，色呈金黄时捞出装盘。

⑤ 锅置火上，入油烧至100℃时，先放入泡辣椒末炒香，再下姜米、蒜米、葱花炒香，然后倒入芡汁，待收汁浓稠、亮油后，起锅淋在藕饼上即成。

风味特色

色泽红亮，外酥里嫩，咸、甜、酸、辣兼备，姜、葱、蒜味突出。

创意设计

霜降时节，四川各地莲藕大量上市。此菜以莲藕为主料，并借鉴川菜名品——“鱼香茄饼”的制法改进而成。在养生方面，莲藕与猪肉搭配，清热生津、润燥益肾，鱼香味型酸甜开胃，最宜此时食用，也可为冬季进补打下基础。藕中含有绿原酸等成分，有抗鼻咽癌的作用。

Ingredients

200g lotus roots; 80g pork; 140ml beaten eggs; 100g cornstarch; 40g pickled chilies, finely chopped; 5g ginger, finely chopped; 15g garlic, finely chopped; 25g spring onions, finely chopped; 3g salt; 10ml Shaoxing cooking wine; 20g sugar; 10ml vinegar; 5ml soy sauce; 3ml sesame oil; 1g MSG; 15g average batter; 100ml stock; 1,500ml cooking oil (about 80ml to be consumed)

Preparation

1. Peel the lotus roots, and cut into connected pairs. Mix 100ml beaten eggs with cornstarch, and blend well to make a paste.
2. Mince the pork, and mix well with salt, cooking wine, beaten eggs to make the filling. Press the filling into lotus root pairs.
3. Mix salt, sugar, cooking wine, vinegar, soy sauce, MSG, batter and stock in a bowl to make the seasoning sauce.
4. Heat oil in a wok to 150℃. Coat the lotus root pairs with the egg-cornstarch paste, deep fry for shaping, and remove. Reheat the oil to 200℃, deep fry the lotus root pairs for a second time till golden brown and crispy on the outside, and transfer to a serving dish.
5. Heat oil in a wok to 100℃, and stir fry the pickled chilies till fragrant. Blend in the ginger, garlic and spring onions, and continue to stir till aromatic. Blend in the seasoning sauce, braise till the sauce thickens and the oil becomes clear, and pour over the lotus root pies.

Features

reddish brown color; crispy pies with soft filling; balanced tastes of saltiness, sweetness, sourness and spiciness,; aromas from garlic, ginger and spring onions

Notes

During the period of Frost's Descent, lotus roots are available in the market. This dish uses lotus roots as the main ingredients, and draws on the cooking method of a famous Sichuan dish called Fish-Flavor Eggplant Pies. In terms of food therapy, the combination of lotus roots and pork helps to remove heat and dryness of the body and enhance the function of the kidney. The sweet and sour tastes of the dish boost the appetite, laying a solid foundation for taking tonics in the coming winter. Besides, the chlorogenic acid in lotus roots prevents nasopharyngeal cancer

木瓜糍粑

Papaya Ciba Rice Cake

食材配方

糯米500克　豆沙馅300克　熟黄豆粉100克　熟白芝麻50克　去皮熟花生50克　木瓜肉250克

制作工艺

① 将熟白芝麻、去皮熟花生磨成粉末，与熟黄豆粉混合均匀备用；将木瓜肉搅打成木瓜汁。

② 糯米淘洗干净后，加入清水浸泡8个小时，之后滤干水分，上笼蒸制成较干的糯米饭，倒出后趁热擂细、制成米团，再加入木瓜汁搅拌均匀待用。

③ 将制好的混合粉铺在案板上，再将木瓜米团摊开，压成厚约0.8厘米的方块，在二分之一处填上豆沙馅后对叠压实，切成菱形块装盘，配上木瓜汁味碟即成。

风味特色

质地软糯，味道甘甜，木瓜香浓。

创意设计

四川地区吃糍粑的习俗历史悠久。此点是在传统糍粑基础上加入木瓜制成，融入了糯米的香甜与木瓜的果香。在养生方面，糯米温中益气，木瓜消食除湿，搭配豆沙祛湿健脾；花生、芝麻与黄豆混合为粉，既香浓又益气润燥。此外，木瓜富含胡萝卜素、木瓜蛋白酶等，能抗菌、抗氧化。

Ingredients

500g glutinous rice; 300g mashed red beans; 100g ground cooked soybeans; 50g cooked sesame seeds; 50g cooked peanuts, peeled; 250g papaya flesh

Preparation

1. Grind the sesame seeds and peanuts, and mix with the ground soybeans. Put the papaya flesh into a blender and blend till smooth.
2. Rinse the glutinous rice, soak in water for eight hours, and drain. Steam in a steamer till just cooked and remove. Hammer and pound the rice to make dough, add papaya juice and knead to mix well.
3. Sprinkle the cutting board with the mixed powder, spread the dough over it and press into a 0.8cm thick square. Smear the red bean paste on half of the square, fold and press. Cut into diamonds, and transfer to a serving dish. Serve with papaya juice dipping sauce.

Features

soft and glutinous in texture; sweet in taste with a strong fragrance of papaya

Notes

The custom of eating Ciba rice cake in Sichuan has a long history. The dish is a modern innovation of traditional Ciba rice cake with papaya added. In term of health benefits, glutinous rice is mild in nature and helps with the movement of qi inside the body, papaya has the function of boosting digestion and removing inner dampness, and red beans are de-humidifying and spleen-invigorating. The aromatic mixture of ground peanuts, sesames and soybeans not only relieves inner dryness but also promotes the flow of qi. Besides, papaya contains carotene and papain, a perfect antidote and antioxidant.

THE BEGINNING OF WINTER

立冬

唐·李白

冻笔新诗懒写，
寒炉美酒对温。
醉看墨花月白，
恍疑雪满前村。

立冬
西风渐作
金叶飘零

立冬物候 | PHENOLOGY IN THE BEGINNING OF WINTER

立冬是二十四节气中的第十九个节气，也是冬季的第一个节气。当太阳到达黄经225° 时为立冬，时间在每年的11月7日~8日。自秦汉以来，我国古人便认为立冬不仅代表着冬天的来临，更意味着万物收藏，以避寒冷。

The Beginning of Winter is the nineteenth of the 24 Solar Terms, and the first solar term in winter. It begins when the sun reaches the celestial longitude of 225°. It usually begins from 7th to 8th November every year. Since Qin Dynasty and Han Dynasty, Chinese people have thought that the Beginning of Winter is not only the start of winter, but also a period when all the living things hide from coldness.

我国古代将立冬后的十五天分为三候：一候水始冰；二候地始冻；三候雉入大水为蜃。立冬初五日，北方天气寒冷，水面上开始结有薄冰，南方正是秋收冬种的好时节；再过五日，土壤中的水分因天冷而凝冻，随之变硬；再经五日，野鸡一类的大鸟就很少见了，禽鸟们或已南迁，或寻一处温暖的地方避寒了。

Ancient China has divided the 15 days after the Beginning of Winter into 3 pentads. In the first pentad, a thin layer of ice forms over the water in the north; and its is the proper season to harvest and plant in the south. In the second pentad, the water inside the soil has been frozen and the soil becomes hard accordingly. In the third pentad, it is hard to find big birds like pheasant; and birds have flown to the south to find a warm place for winter.

食材生产 | PRODUCING FOOD INGREDIENTS

立冬前后，适逢秋、冬两季的更替阶段，冷空气带来的更多是大风降温。此时，四川大部分地区的气候仍然温暖宜人，农谚说，“十月小阳春，无雨暖烘烘”，正是油菜、小麦、豆类查苗、补苗、移栽的时候，各地把田间管理放在首位，同时，四川地区麦苗青翠、豆苗茁壮。立冬以后，也是“小秋收”的黄金季节，要做好各类果实种子的采收与储藏。冬季果树处于休眠期，可进行修枝、剪枝，以减少病虫害，提高果树来年的结果质量。

Around the Beginning of Winter, it is the turn of autumn and winter. The cold air will bring high wind and falling temperature to the land. At this moment, it is warm in most parts of Sichuan. As the farmer’s proverb goes, “The weather of late autumn in October is warm and sunny. ” It is the right time to transplant rape and wheat, check the seedlings of beans, and fill the gaps with seedlings. Fields management is the priority. Meanwhile, the wheat seedling is green and the bean seedling is strong in Sichuan. After the Beginning of Winter, it is the gold season for minor autumn harvest. People should pick and store well all kinds of fruits and seeds. Fruit trees will hibernate in winter. Then people can prune them to reduce pest and improve the quality of harvest next year.

立冬作为“四时八节”之一，在我国古代备受重视，常常要举行一系列的祭祀活动，其主要目的是为了祈求上苍保佑，能够让来年风调雨顺、五谷丰登，过上丰衣足食的生活。四川民间在立冬时节有用糍粑祭祀耕牛的民俗活动，民国二十三年（1934年）的《华阳县志》，就记载了当时立冬时节盛行的“牛王会”，“是日为牛王生日，乡农以糍粑置牛角相庆贺”。《三台县志》载，“农家做糍糕、饭牛，复黏角上，令其临水照，则牛喜，又名接牛角”。耕牛是农业文明的象征，祭牛旨在祈求来年五谷丰登。

As one of the four seasons and eight solar terms, the Beginning of Winter have carried great weight in ancient China. There normally were a series of ritual activities for good weather for the crops and a golden harvest next year. The date of the Beginning of Winter is close to Ox King Festival. On this day, Sichuan people will offer Ciba rice cake to farm cattle. *Huayang County Chronicles* in the 23rd year of the Republic of China recorded, “that day is the Ox King’s birthday, and people put Ciba rice cake on the horns for celebration.” *Santai County Chronicles* recorded, “farmers make Ciba rice cake for cattle, and put the cake on horns which is called extension of horn. Cattle will watch its reflection in water and feel happy. ” Farm cattle is the symbol of agriculture civilization. Offering sacrifice to ox is for golden harvest in the coming year.

饮食养生 | DIETARY REGIMEN

立冬象征着秋去冬来，天气由凉转寒。冬季是一年中最为寒冷的季节，寒风阵阵，草木凋零，蛰虫伏藏，阴盛阳衰，人体的新陈代谢也较缓慢，需要养精蓄锐、安度隆冬，为来年春天生机勃发做好准备，是人体“养藏”的最佳时机。在四川地区，除了川西高原阳光充足外，其他地方少见阳光普照，冬季尤显阴冷，因此更需注意保护阳气。“冬令进补”是民间习俗，也是我国传统医学和养生学用以治疗虚弱病、改善体质、提高机体免疫力和延年益寿的有效手段。立冬可多吃蔬菜和鸡、甲鱼、羊肉、桂圆、木耳等，也可适当增加动物内脏、瘦肉、鱼类、蛋类等。若非阴虚及燥热体质，可适量温补，如选用羊肉、牛肉、鳝鱼等。这些食材不但味美，而且富含蛋白质、脂肪、碳水化合物，以及钙、磷、铁等多种营养成分，能补充因冬季寒冷而消耗的热量，还能益气、养血、补虚。同时，在饮食上还应遵循“省咸增苦”的原则，避免过量食用咸味食物，最好能适量吃一些苦味食物，如青菜、青菜头、瓢儿白、油菜等略带苦味的应季时蔬。初冬时节是心血管病的高发期，可多吃清润甘酸的食物，如适当多吃些醋，能软化血管、预防心血管疾病的发生。

The Beginning of Winter symbolizes the departure of autumn and the arrival of winter. The weather has changed from cool to cold. Winter is the coldest season in a year. The wind is piercingly cold; the plants have withered; and insects have hibernated underground. In winter, the metabolism of human body is slow and yin rises while yang declines, hence people should conserve strength and store up energy for vigorous renewal in the coming spring. It is the best time for “nourishing and storing our energy”. In Sichuan region, sunshine is rare and winter is gloomy in most places except for Western Sichuan Plateau. Therefore, local people have to pay more attention to protecting yang qi. “Tonify human body in winter” is a folk custom; and it is an efficient way to cure weakness, improve physique, enhance immunity and prolong life according to TCM and regimen. In the Beginning of Winter, people should eat more vegetable, chicken, soft-shelled turtle, mutton, longan, agaric, and more animal innards, lean meat, fish, egg, etc. If you are not the type of person in deficiency of yin and interior heat, you can warmly tonify the body by eating mutton, beef, earl properly. Those food ingredients are not only delicious, but also full of protein, fat, carbohydrate, calcium, phosphorus, iron element, which can replenish the energy lost in winter, and tonify blood and qi. Moreover, people should follow the principle of “less salty and more bitter” when eating. We should avoid too many salty foods, and

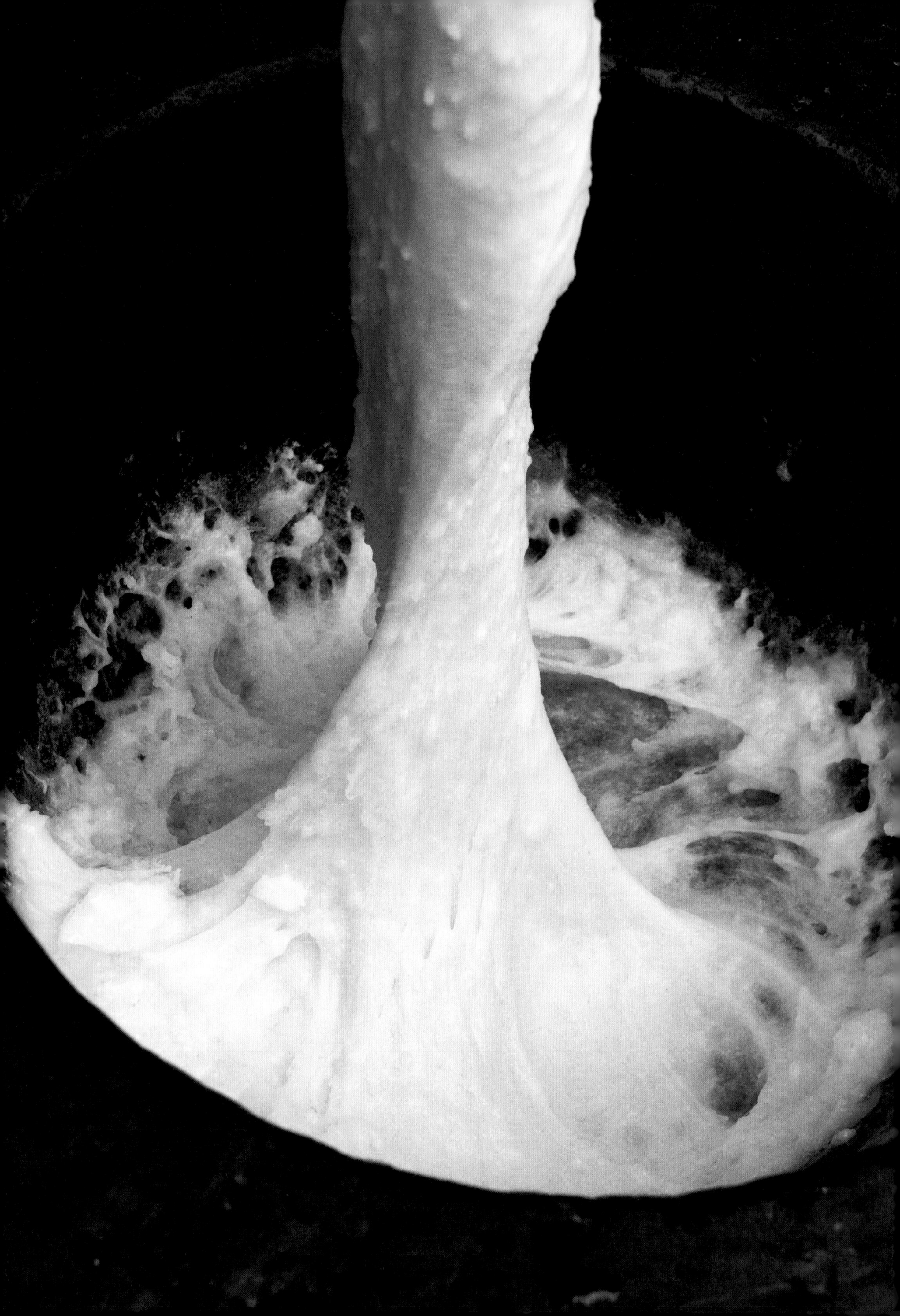

take more bitter foods like Chinese cabbage, cabbage head, bok choy, rape. Early winter is the cardiovascular diseases season. People should take more light, moist, sweet and sour foods to soften the blood vessels and prevent cardiovascular diseases, for example, vinegar.

立冬美食 | FINE FOODS IN THE BEGINNING OF WINTER

立冬的主要食俗有三项：一是“补冬”。俗语言，“立冬补冬，补嘴空”。家家户户杀鸡宰鹅，准备猪肉、牛肉和鱼虾，以滋补身体。二是腌菜。立冬时节正是制作腌菜的好时机，人们把初熟的新鲜蔬菜加以腌藏，以备冬日之需。据宋代孟元老的《东京梦华录》记载：“是月立冬，前五日，西御园进冬菜，上至官禁，下至民间，一时收藏，以充一冬食用。”在四川地区家喻户晓的“老坛酸菜”，就常在此时用大叶青菜发酵腌制。三是打糍粑。四川各地乡间多有“十月初一打糍粑”的风俗，糍粑是用当年收获的新糯米经蒸熟、捣烂后制成，再将芝麻、黄豆或花生炒香后磨粉，加糖而食，香甜软糯，意味着丰收的喜悦。

There are three major eating customs in the Beginning of Winter. Firstly, it is “nourishing the body in winter”. As the proverb goes, “people should nourish the body and fill the mouth in the Beginning of Winter ”. Every family will kill chicken and goose, and prepare pork, beef, fish, shrimp to nourish the body. Secondly, it is making pickles. The Beginning of Winter is the best time to make pickles. People will store the fresh vegetables with salt for the coming winter days. Meng Yuanlao in Song Dynasty (960-1279) has recorded in his *Records of Dreamily Prosperous Eastern Capital*, “5 days before the Beginning of Winter, western royal court will prepare preserved vegetables. From the emperor to the common people, everybody will eat them in whole winter.” The famous Sichuan Pickled Chinese Cabbage are made of salty Chinese cabbage with big leaves just in this time. Thirdly, it is beating Ciba rice cake. There has been a custom of “beating Ciba rice cake on the first day of October” in the countryside of Sichuan Province. Ciba rice cake is made by steaming and mashing glutinous rice newly harvested in that year; then eaten with sugar and ground sesame, soybean or peanut which are fried till fragrant. The sweet and glutinous Ciba rice cake means the joy of harvest.

一品牦牛掌

Yipin Yak Paw

食材配方

牦牛掌1个（约1000克） 土母鸡肉500克 猪肘肉500克 熟火腿200克 白萝卜500克 瓢儿白150克 郫县豆瓣50克 花椒3克 姜片30克 葱段50克 食盐5克 酱油10毫升 料酒50毫升 味精2克 胡椒粉1克 芝麻油3毫升 花椒油10毫升 鲜汤2000毫升 水淀粉15克 食用油80毫升

制作工艺

① 牦牛掌初加工后用清水多次焯水，入锅煮2小时至牦牛掌变软后捞出，剔除掌骨，再入锅内，加鲜汤、姜片10克、葱段20克、料酒煨煮1小时，此后重复两次去除膻味。

② 将牦牛掌横切数刀，以不切穿掌底为度，放入纱布袋中包好定型；鸡肉、猪肘肉分别切块后焯水；白萝卜用模具压成厚约2厘米的圆形快，焯水备用；瓢儿白一破为二，焯水备用。

③ 锅置火上，入油烧至120℃时，放入郫县豆瓣、姜片、葱段、花椒炒香，掺入鲜汤，放入鸡块、猪肘、火腿、食盐、酱油、料酒、胡椒粉和牦牛掌包，用小火烧2～3小时至熟软，再放入白萝卜烧至熟软，然后捞出牦牛掌，在盘中摆为“一品形”，将白萝卜摆在四周，点缀上飘儿白。

④ 在锅中汤汁中加入味精、芝麻油、花椒油、水淀粉收汁至浓稠，起锅浇淋在牦牛掌上即成。

风味特色

成菜色泽红亮，掌形完整，肉质软糯，香味浓郁，味道咸鲜，略带麻辣。

创意设计

牦牛掌是四川特产的一种滋补食材，此菜以牦牛掌为主要食材制成，口感咸鲜，略带麻辣的家常风味比较突出。在养生方面，立冬之时是进补之始，牦牛掌性温，可补脾气、强筋骨、美肌肤，辅以鸡肉、猪肘及火腿，则能起补血益精、滋阴润燥等作用，且都是高蛋白、低脂肪之物，营养价值高；郫县豆瓣、花椒、姜虽辛散，但煨炖成菜则能减缓其燥性；瓢儿白味略清苦，不仅解腻，也符合冬季“增苦”的养身理念。整道菜既有立冬温补之功，又不会耗散太过。

Ingredients

1 yak paw(about 1,000g); 500g free-range hen; 500g pork knuckle; 200g precooked ham; 500g white radishes; 150g bok choy; 50g Pixian chili bean paste; 3g Sichuan peppercorns; 30g ginger, sliced; 50g spring onions, segmented; 5g salt; 10ml soy sauce; 50ml Shaoxing cooking wine; 2g MSG; 1g pepper; 3ml sesame oil; 10ml Sichuan pepper oil; 2,000ml stock; 15g average batter; 80ml cooking oil

Preparation

1. Blanch the cleaned yak paw in water several times, and boil for about 2 hours till soft, then remove and de-bone. Put the paw into a wok, add some stock, 10g ginger, 20g spring onions and cooking wine, and bring over a low flame to a boil and simmer for about 1 hours. Repeat the simmering process two more times to remove the unpleasant smell of the paw.
2. Make several cuts into the paw and wrap them up in cheesecloth. Chunk the hen and pork knuckle, then blanch separately. Mould the radishes into 2cm thick round shapes, and blanch. Halve the bok choy, and blanch.
3. Heat oil in a wok to 120℃, add the Pixian chili bean paste, ginger, spring onions and Sichuan peppercorns, and stir-fry till aromatic. Add stock, chicken, pork knuckle, ham, salt, soy sauce, cooking wine, pepper and wrapped paw, and braise over a low flame for 2 to 3 hours till soft and cooked through. Then, add radishes and braise till soft. Remove the paw from the wok, and lay neatly on a serving plate. Arrange the radishes around the paw, and garnish with bok choy.
4. Add MSG, sesame oil, Sichuan pepper oil and batter to the wok, simmer till the soup thickens, and pour over the paw.

Features

bright and lustrous color; glutinous and tender texture; savory, succulent and slightly spicy tastes

Notes

People usually start to take tonics when winter comes. Yak paw, a Sichuan specialty, is a nourishing health ingredient. In food therapy, Yak paw has warm nature, which can tonify the spleen, strengthen the sinews and beauty the skin. Combining chicken, white radishes and ham, this dish can tonify and replenish blood, nourish yin and moisten dryness. Besides, all these ingredients are high-protein and low-fat foods with rich nutrients. Although chili bean paste, Sichuan peppercorns and ginger are pungent, braising can reduce their dry nature. The bitter taste of bok choy can reduce the greasiness of the dish, and is accorded with the body demand for more bitterness during the winter. The dish on the whole is a mild tonic during the Beginning of Winter.

冬菜扣肉

Dongcai Kourou

食材配方

带皮猪五花肉250克　四川冬菜200克　泡辣椒10克　姜片5克　葱15克　豆豉10克　食盐2.5克　胡椒粉0.5克　味精1克　料酒8毫升　酱油10毫升　冰糖糖色10克　食用油1000毫升（约耗20毫升）

制作工艺

① 将猪五花肉放入冷水锅内煮至断生后捞出，搌干水分，趁热在表皮抹上糖色；四川冬菜洗净，切碎。

② 锅置火上，入油烧至180℃时，放入猪肉炸至表皮起皱，色呈棕红时捞出，再放入煮肉的汤中浸泡至温热后捞出，切成长约10厘米、宽约5厘米、厚约0.3厘米的片。

③ 食盐、酱油、料酒、味精、冰糖糖色、胡椒粉入碗调成味汁。

④ 将猪肉片的皮向下摆入蒸碗中呈"一封书"造型，再放入冬菜、泡辣椒、葱、姜片、豆豉，淋入味汁，入笼蒸至软熟后取出，翻扣入盘内即成。

风味特色

成菜色泽棕红，质地炬软，咸鲜醇香，肥而不腻。

创意设计

扣肉是四川资阳等地民间"三蒸九扣"的代表菜品之一，因将肉片翻扣成菜而得名。冬菜是四川四大腌菜之一，此菜以猪肉、冬菜为食材制作而成，极具四川特色。在养生方面，猪肉补虚、滋阴、润燥；冬菜通利肠胃、养胃和中，二者搭配，既滋阴养胃，又缓解油腻，还可补充膳食纤维和钙、铁、锌等微量元素，营养价值较高。

Ingredients

250g pork belly with skin attached; 200g Sichuan dongcai (preserved cabbage); 10g pickled chili, chopped diagonally; 5g ginger, sliced; 15g spring onions, chopped diagonally; 10g fermented soybeans; 2.5g salt; 0.5g pepper; 1g MSG; 8ml Shaoxing cooking wine; 10ml soy sauce; 10g caramel; 1,000ml cooking oil (about 20ml to be consumed)

Preparation

1. Add the pork belly to a pot with cold water, and boil until just cooked, then ladle out and drain. Brush with caramel while it is still warm. Rinse and chop the dongcai.
2. Heat oil in a wok to 180℃, add the pork to deep fry till the skin is wrinkled and dark brown. Ladle out and soak in the water where the pork have been boiled. Wait till the pork is warm, remove and cut into 10cm long, 5cm wide and 0.3cm thick slices.
3. Blend salt, soy sauce, cooking wine, MSG, caramel and pepper in a bowl to make the seasoning sauce.
4. Lay in a steaming bowl the pork slices with the skin downwards, presenting a form of a book. Add dongcai, pickled chilies, spring onions, ginger, fermented soybeans, and drizzle with the seasoning sauce. Steam the pork till fully cooked and remove. Turn the bowl upside down to transfer the contents onto a serving dish.

Features

brown color; soft and glutinous pork that is fatty but not greasy; delicate, aromatic and savoury tastes.

Notes

Kourou, reversion pork if translated literally, is a representative of Ziyang traditional dishes. It is so named because the steaming bowl is reversed onto the serving dish. Dongcai is one of the "Four Famous Preserved Vegetables" in Sichuan. In food therapy, pork has the function of promoting health, nourishing yin and moistening dryness. Dongcai can nourish the stomach and stimulate peristalsis. Combining the two ingredients can not only replenish yin and benefit the stomach, but reduce the greasiness and provide dietary fiber, calcium, iron and zinc for human body.

菌香冬饺

Winter Jiaozi with Mushrooms

食材配方

面粉250克　清水120毫升　猪肉末200克　鸡纵菌150克　食盐3克
味精1克　胡椒粉0.5克　芝麻油3毫升　酱油20毫升　姜葱水10毫升
鲜汤40毫升　白糖1克　蒜泥5克　红油10毫升　葱花1克

制作工艺

① 面粉加冷水和成冷水面团，盖上湿毛巾醒面30分钟。把鸡纵菌切成小块备用。
② 猪肉末加食盐、味精、胡椒粉、芝麻油、酱油、姜葱水、鲜汤拌匀，再加入切碎的鸡纵菌颗粒，拌匀成馅料。
③ 将面团搓成长条，揪成每个重约8克的剂子，再擀成中间厚，边缘薄，直径约7厘米的圆皮，包入馅料，捏成饺子生坯。
④ 锅置火上，加水烧沸，下饺子生坯煮至成熟后捞出盛入碗中；将酱油、白糖、蒜泥、红油、葱花调制成红油味碟，与饺子同上即成。

风味特色

皮薄馅嫩，筋道爽滑，菌香突出，味道咸鲜，略带甜辣。

创意设计

俗话说“立冬不端饺子碗，冻掉耳朵没人管”，我国自古以来，就有在立冬日吃饺子的习俗。菌香冬饺采用资阳等地出产的鸡纵菌为馅心制成，地方特色浓郁。在养生方面，鸡纵菌益胃健脾、清爽神志，与猪肉、面粉搭配做成饺子，更能益气温阳。此外，鸡纵菌富含膳食纤维和多糖，对抗癌、抗衰老有一定的辅助作用。

Ingredients

250g wheat flour; 120ml water; 200g pork mince; 150g jizong mushrooms; 3g salt; 1g MSG; 0.5g pepper; 3ml sesame oil; 20ml soy sauce; 10ml ginger and spring onion juice; 40ml stock; 1g sugar; 5g garlic, finely chopped; 10ml chili oil; 1g spring onions, finely chopped

Preparation

1. Mix the wheat flour with cold water to make dough, cover with a wet towel and let rest for 30 minutes. Chop the jizong mushrooms into cubes.
2. Mix the pork mince with salt, MSG, pepper, sesame oil, soy sauce, ginger and spring onion juice and stock. Then add the jizong mushrooms and mix well to make the stuffing.
3. Knead the dough into a long strip, tear into 8g equal portions. Flatten them with a rolling pin to make round, middle-thick and edge-thin wrappers, 7cm in diameter. Put some stuffing on a wrapper, fold the wrapper to wrap the stuffing up. Press hard and secure the edges to make jiaozi.
4. Heat water in a wok to bring a boil, add jiaozi, and boil till cooked through. Ladle out the jiaozi to a serving bowl. Blend soy sauce, sugar, garlic, chili oil and spring onions to make the dipping sauce, and serve with jiaozi.

Features

thin wrappers and tender stuffing; smooth and springy bite; savoury, slightly sweet and spicy tastes; strong mushroom aroma

Notes

As the old saying goes, eating jiaozi in the Beginning of the Winter does not freeze the ears. Chinese people have had a tradition of eating jiaozi on the Beginning of Winter Day since ancient time. Jiaozi in this dish uses the jizong mushrooms in Ziyang to make the stuffing. In food therapy, jizong mushrooms can strengthen the spleen and the stomach and refresh the mind. Making dumplings with this mushroom and pork can replenish qi and warm yang. Besides, jizong mushrooms have rich dietary fiber and polysaccharide, which have anti-cancer and anti-aging properties.

LESSER SNOW

小雪

花雪随风片片玲珑

小雪

宋·陆游

檐飞数片雪，
瓶插一枝梅。
童子敲清磬，
先生入定回。

小雪物候 | PHENOLOGY IN LESSER SNOW

小雪是二十四节气中的第二十个节气。当太阳到达黄经240° 时为小雪，时间在每年的11月22日~23日。小雪节气的到来，标志着气温明显走低，降雪逐渐开始。农谚说，“小雪雪花飞”，这在黄河流域及以北地区表现极为明显。

Lesser Snow is the twentieth of the 24 Solar Terms. It begins when the sun reaches the celestial longitude of 240°. It usually begins from 22nd to 23rd November every year. The arrival of Lesser Snow means that the temperature will drop significantly and it is going to snow. As the farmer's proverb goes, "It is snowing in Lesser Snow". This kind of weather is very obvious in Yellow River Basin and the north of the Yellow River.

我国古代将小雪后的十五天分为三候：一候虹藏不见；二候天气上升，地气下降；三候闭塞而成冬。小雪时节，由于气温较低，北方下雪逐渐增多，所以彩虹就像藏起来一样，看不见了；天空中的阳气上升，地中的阴气下降，导致天地不通、阴阳不交；万物失去生机，天地闭塞而转入严寒的冬天。

Ancient China has divided the 15 days after Lesser Snow into 3 pentads. in the first pentad, snow has increased in the north because of low temperature; and people can hardly see rainbows. In the second pentad, yang qi in air is rising, and yin qi on the ground is descending. In the third pentad, all things on earth have lost vitality, and the whole world enters the closed winter.

食材生产 | PRODUCING FOOD INGREDIENTS

小雪时节，四川盆地北部及高原地区率先进入冬季，“荷尽已无擎雨盖，菊残犹有傲霜枝”的初冬景象随处可见。因为北有秦岭、大巴山为屏障，阻挡了部分冷空气的入侵，在一定程度上削减了寒潮的威势，因此四川盆地少有降雪，多以潮湿、阴冷的天气为主。资阳地处四川盆地中部，小雪前后的平均气温在11℃~17℃，与初雪降临、安静闲适的北方大地不同，资阳地区广袤的农村依然生机盎然，农人们忙于采收果园中的柑橘、柠檬等水果，还要在菜地里种上蚕豆、豌豆、莲花白、莴笋、棒菜、儿菜等，以便严寒的冬天也能享用到新鲜的蔬菜。

In Lesser Snow, the north of Sichuan Baisn and plateau enter the winter firstly. Lotus blossoms and leaves are no more, and the rest of chrysanthemums are still on the branches with frost. The Qinling Mountains in the north and Daba Mountains in the south have blocked part cold air, and weakened the power of cold wave. It rarely snows in Sichuan Basin. The weather is humid and gloomy there. Ziyang is in the center of Sichuan Basin, the average temperature is from 11℃ to 17℃ around Lesser Snow. Unlike first-snowing in quiet northern region, the vast rural area in Ziyang region is bustling. Farmers are busy with picking citrus, lemon in orchard, and plant broad bean, pea, cabbage, lettuce, mustard, etc. for the coming severe winter.

饮食养生 | DIETARY REGIMEN

我国传统医学和养生学认为，阳光对人的照射能够起到助发人体阳气的作用，非常有益健康。但是，小雪时节气候寒冷、潮湿，大自然处于天光晦暗的阴盛阳衰状态，而天人相应，人体也必然处于阴冷、抑郁的阳衰阶段，心情也会受到影响。小雪节气的养生需要遵从冬季益肾藏精、敛阴护阳的总原则，在饮食上除选用羊肉、牛肉等温补食材外，还宜常食栗子、腰果、核桃等坚果，以补脾、补肾。此外，白菜、萝卜、芹菜等应季时蔬也可经常食用，而兔肉、黑木耳、紫甘蓝、山楂能活血化瘀，对冬季预防心脑血管疾病也有一定帮助。

According to TCM and regimen, the sunshine can help the growth of yang qi in human body, and it is very important for health. However, it is cold and humid in Lesser Snow. The yin is strong and yang is weak in nature. Human body is also in a stage of gloom and weak yang. In Lesser Snow, people should follow the principle of protecting kidney and storing essence, protecting yang and reducing yin. People should eat more tonic foods like mutton and beef, and eat chestnut, cashew nut, walnut to tonify spleen and kidney. Furthermore, people should eat more seasonable vegetables, like Chinese cabbage, radish, celery. Rabbit meat, black fungus, purple cabbage, haw, can promote blood circulation to remove blood stasis, and can help to prevent cardiovascular diseases in winter.

小雪美食 | FINE FOODS IN LESSER SNOW

“冬腊风腌，蓄以御冬。”小雪一到，气温急剧下降，北风凛冽，正是加工腊肉的好时机，资阳乃至四川大部分地区的家庭开始腌腊肉、灌香肠。将猪肉切割后，加入一定比例的花椒、大料、桂皮、丁香等香料腌渍，用绳串挂起来，或选用柏树枝、甘蔗皮、柴草等慢慢熏烤，或挂于农家灶头顶上，利用烟火慢慢熏干。制作好的腊肉、香肠色泽鲜艳、味道醇香、营养丰富，且具有开胃、祛寒等功能。如今，小雪以后，在四川地区许多家庭的阳台、窗外都会挂起串串香肠、腊肉，业已成为冬季一道独特的美食风景。

“Preserve foods for winter”. With the arrival of Lesser Snow, the temperature drops quickly and brisk wind blows. It is the right time to make Chinese bacon. In most parts of Sichuan including Ziyang, people start to make Chinese bacon and sausage. Farmers slice the pork first, and then preserve it with proportional Sichuan pepper, aniseed, cinnamon, clove. Later, farmers will hang those pork and fumigate them by burning cypress, sugarcane rind, firewood, or hang those pork above the farmer’s kitchen range directly. Those Chinese bacon and sausage are good-looking, delicious, nutritious, and have the function of whetting appetite and dispelling coldness. Nowadays, lots of families in Sichuan will hang many Chinese bacons and sausages out of the window or in the balcony after Lesser Snow, which has become a special landscape in winter

生爆甲鱼

Stir Fired Softshell Turtle

食材配方

甲鱼1只（约800克） 青红辣椒各200克 独蒜200克 姜片20克 青花椒粒30克 郫县豆瓣40克 白糖2克 味精2克 胡椒粉1克 酱油5毫升 料酒15毫升 蚝油10毫升 藤椒油10毫升 鲜汤200毫升 食用油1000毫升（约耗120毫升）

制作工艺

① 甲鱼宰杀、治净，放入70～80℃的热水中烫皮，褪去表面薄膜后斩成约3厘米见方的块。

② 青红辣椒分别切成约1.5厘米大的丁；独蒜切成大丁。

③ 锅置火上，入油烧至180℃时，放入甲鱼块炸至成熟，沥去余油后放入郫县豆瓣炒香，接着下蒜丁、姜片、青花椒粒炒香，再入鲜汤、白糖、胡椒粉、酱油、料酒、蚝油，烧至汁水将干时，入青红辣椒、味精、藤椒油焖1分钟，起锅装盘即成。

风味特色

成菜色泽红亮，质地软糯，味道咸鲜麻辣。

创意设计

俗语说“春夏养阳，秋冬养阴”。小雪时节寒气加重，人体代谢功能下降，急需进补一些滋阴润燥之品以防伤阴。此菜以甲鱼为主要食材制成，是应季的食用佳品。甲鱼滋阴补肾、清退虚热，富含17种氨基酸及10余种微量元素，养阴之功显著。再辅之以辣椒、花椒、姜、蒜等辛香开胃之品，既充分体现出川菜的风味特色，又能温通开胃，避免滋腻，食疗价值显著。

Ingredients

1 softshell turtle (about 800g); 200g green chili peppers; 200g red chili peppers; 200g garlic, sliced; 20g ginger; 30g green Sichuan peppercorns; 40g Pixian chili bean paste; 2g sugar; 2g MSG; 1g pepper; 5ml soy sauce; 15ml Shaoxing cooking wine; 10ml oyster sauce; 10ml green Sichuan pepper oil; 200ml stock; 1,000ml cooking oil (about 120ml to be consumed)

Preparation

1. Kill the turtle and rinse. Blanch in 70℃ to 80℃ hot water to remove the membrane, then chop into 3cm cubes.
2. Chop the green and red chili peppers into 1.5cm cubes separately. Chop the garlic into big cubes.
3. Heat oil in a wok to 180℃, fry the turtle till cooked through, then remove. Pour out extra oil, add Pixian chili bean paste to stir fry till aromatic, and add garlic, ginger, green Sichuan peppercorns to stir fry till aromatic. Blend in the stock, sugar, pepper, soy sauce, cooking wine and oyster sauce, continue to braise to reduce the sauce. Add the red and green peppers, MSG and green Sichuan pepper oil, cover and simmer for one more minute. Transfer to a serving dish.

Features

reddish brown color; tender and glutinous turtle meat; spicy and savoury tastes

Notes

Here goes a saying, “Spring and summer are the best time to nourish yang while autumn and winter are the best time to nourish yin”. Because of the colder weather and lower human body’s metabolism, it is necessary to have food which nourish yin and moisturize dryness of human body during Lesser Snow period. This dish adopts softshell turtle as the main ingredient, which is rich in seventeen kinds of amino acids and more than ten sorts of microelements. It is an excellent food to nourish yin, clear heat and tonify the kidney. With the combination of pungent pepper and ginger, this dish has strong features of Sichuan cuisine. It has a high value in food therapy, for it gently unclogs the body and avoids grease accumulation.

素炒野鸡红

Stir Fried Pheasant Tail

食材配方

胡萝卜200克　芹菜100克　蒜苗50克　食盐4克　鸡精1克　味精1克　白糖1克　化猪油50毫升

制作工艺

① 胡萝卜去皮，切成长约10厘米、粗约0.2厘米长的丝；芹菜、蒜苗切成长约7厘米的粗丝。

② 锅置火上，入油烧至160℃时，放入胡萝卜丝翻炒断生，再入芹菜、蒜苗、食盐、鸡精、味精、白糖炒香，起锅装盘即成。

风味特色

成菜色泽鲜艳，味道咸鲜清香。

创意设计

炒野鸡红是川菜著名的一道传统大众家常菜，常用肉丝和胡萝卜、芹菜、蒜苗混合炒制而成，因色彩斑斓，形似野鸡的羽毛而得名。此菜去掉肉丝，将多种蔬菜素炒而成，色泽更鲜艳，滋味更清香，食疗价值独特。胡萝卜素称“小人参”，能健脾和中、滋肝明目，芹菜平肝清热，蒜苗凉血散血，富含维生素、膳食纤维、类胡萝卜素及花青素，在食肉较多的时节食用，更利于疏通肠胃，降低血糖、血压和胆固醇。

Ingredients

200g carrots; 100 celery; 50g baby leeks; 4g salt; 1g chicken essence; 1g MSG; 1g sugar; 50ml lard

Preparation

1. Peel the carrots, then cut into 10cm long and 0.2cm wide slices. Cut the celery and baby leeks into 7cm long thick slivers.
2. Heat oil in a wok to 160℃, add carrots to stir fry till al dente. Add celery, baby leeks, salt, chicken essence, MSG, sugar to stir fry till aromatic, then remove and transfer to a serving dish.

Features

colorful look; savory taste and fresh aroma

Notes

Stir Fried Pheasant Tail is a famous traditional household Sichuan dish, using a variety of vegetables as the main ingredients. It is so named because the colorful vegetables look like pheasant tails which are also of varied colors. A stir fry with carrots, celery and baby leeks as the main ingredients, this dish has fresh aroma and unique health function. In food therapy, carrots are often called “lesser ginseng”, which can fortify the spleen and replenish qi, enrich the liver and improve vision. Celery can pacify the liver and clear heat. Baby leeks can cool the blood and has rich vitamins, dietary fiber, carotenoid and anthocyanidin. It is much more helpful to have this dish when having a lot of meat, for it can soothe bowel system, lower the blood sugar, alleviate the blood pressure and ease cholesterol.

麻香包谷饼

Sesame-Stuffed Corn Pies

食材配方

中筋面粉400克　玉米粉100克　黑芝麻馅200克　安琪酵母5克
无矾泡打粉3克　白糖50克　清水260毫升　化猪油20毫升
去皮白芝麻100克　食用油1000毫升（约耗30毫升）

制作工艺

① 面粉、玉米粉混合均匀，加入酵母、无矾泡打粉、白糖、清水调制成面团，再加入化猪油揉匀，盖上湿毛巾醒面5分钟。

② 将黑芝麻馅搓成每个重约20克的圆球馅心待用。

③ 将面团搓成长条，先揪成每个重约30克的剂子，擀成中间厚、边缘薄的圆片，包入黑芝麻馅心，再捏成圆形包谷饼生坯，在表面刷上清水，粘上白芝麻，放入刷有油脂的蒸笼内待其充分发酵。

④ 锅置火上，入食用油烧热，将包谷饼生坯放入油锅中炸至色金黄时捞出即成。

风味特色

饼色金黄，外酥内软，味道甘甜，芝麻香浓。

创意设计

“包谷”是玉米在四川方言中的别称。麻香包谷饼以资阳所产玉米粉与面粉混合创新而成，粗细搭配，营养均衡。在养生方面，面粉温中益气，玉米调中开胃，辅以黑、白芝麻，更能补气益精，其所含的大量不饱和脂肪酸能降血糖、降血压及抗衰老。其中，黑入肾，黑芝麻尤其适合冬季食用。

Ingredients

400g all-purpose flour; 100g corn flour; 200g black sesames filling; 5g Angel yeast; 3g alum-free baking powder; 50g sugar; 260ml water; 20ml lard; 100g white sesame, peeled; 1,000ml cooking oil (about 30ml to be consumed)

Preparation

1. Mix the all-purpose flour and corn flour, add yeast, alum-free baking powder, sugar and water to make dough. Add lard and mix evenly. Cover the dough with a wet towel and let stand for 5 minutes to leaven.

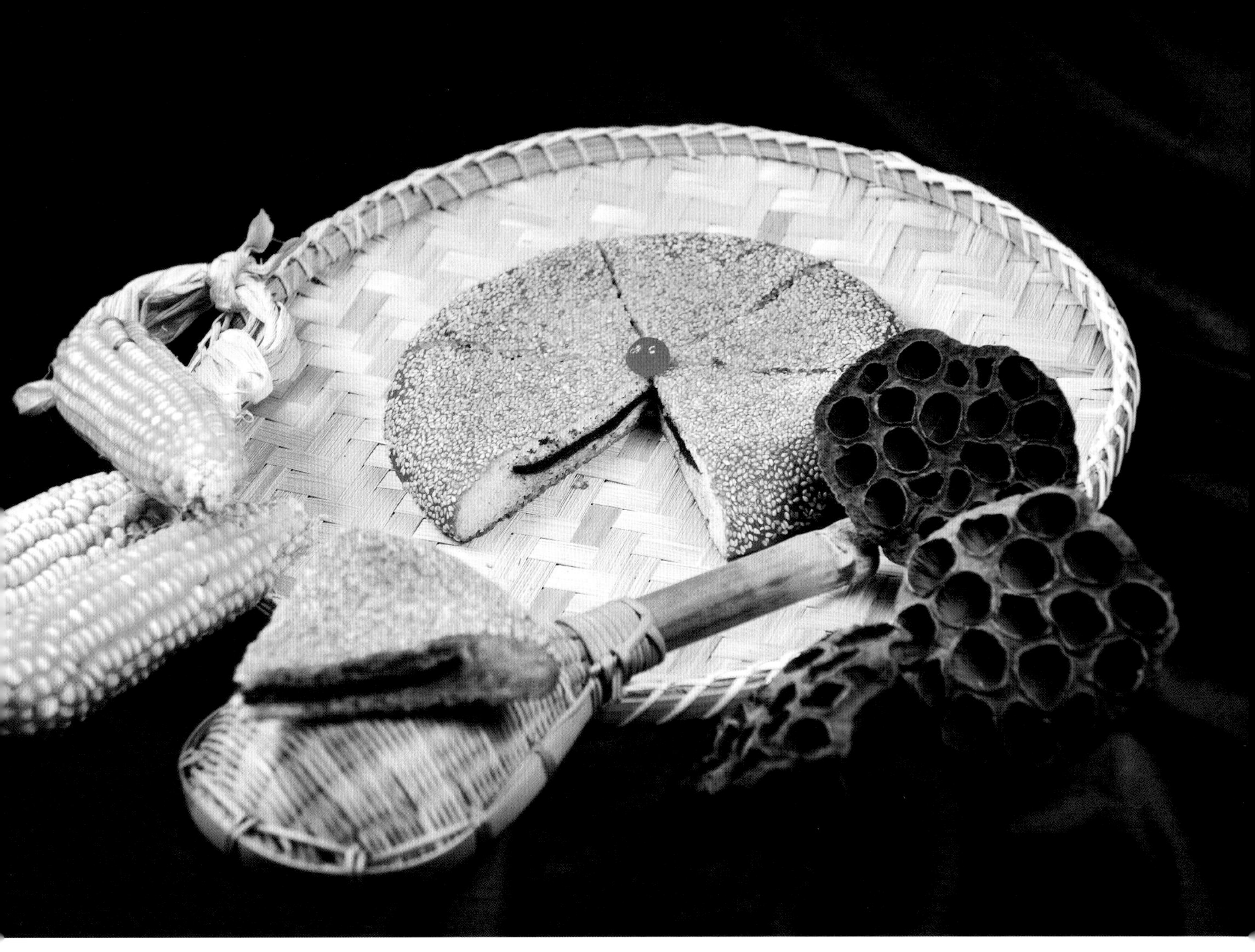

2. Knead the black sesame filling into balls about 20g each.
3. Knead and roll the dough into a log, cut into sections about 30g each and then press to shape each section into a round wrapper which is thick in the middle and thin on the edges. Put some filling on the center of the wrapper. Wrap up and seal. Knead and shape into round shapes. Brush water on the surface of the round dough pieces and coat with white sesames. Transfer the dough to a steaming rack smeared with oil, and wait till it is fully leavened.
4. Heat oil in a wok and deep fry the dough till golden brown.

Features

golden brown color; flaky and crunchy crust; tender stuffing; sweet taste; strong sesame flavor

Notes

The Sesame-Stuffed Corn Pies is made of corn flour produced in Ziyang with all-purpose flour. Using both whole grains and refined grains, the dish is well-balanced in nutrition. In food therapy, wheat flour is mild in nature and benefits qi. Corn helps to stimulate appetites. Combined with black and white sesames, the dish has a stronger function of invigorating the body. Meanwhile, the dish is rich in unsaturated fatty acid, which has anti-aging effect and helps to lower the blood glucose and blood pressure. Among the ingredients, black sesames are good for nourishing the kidney, most suitable for winter.

GREATER SNOW

大雪

北风卷地万山凋敝

大雪书怀

宋·范成大

天将奇赏发清欢，
畴昔登临插羽翰。
梅下寻诗骑马滑，
松梢索酒倚楼寒。
闭门老子愁无赖，
返棹归来兴已阑。
聊掬玉尘添石鼎，
自煎鱼眼破龙团。

大雪物候 | PHENOLOGY IN GREATER SNOW

大雪是二十四节气中的第二十一个节气。当太阳到达黄经255° 时为大雪，时间为每年的12月6日～8日。元朝吴澄《月令七十二候集解》言："大雪，十一月节。大者，盛也。至此而雪盛矣。"大雪时节，往往在强冷空气前沿冷暖空气交锋的地区，会普降大雪甚至暴雪，它和小雪、雨水、谷雨等节气一样，常常是直接反映降水量的节气。

Greater Snow is the twenty-first of the 24 Solar Terms. It begins when the sun reaches the celestial longitude of 255°. It usually begins from 6th to 8th December every year. In his book *A Collective Interpretation of the Seventy-Two Phenological Terms*, Wu Cheng (Yuan Dynasty 1271-1368) said, "the Greater Snow is the November (lunar calendar) Festival. It snows heavily in this period". In Greater Snow, we will experience heavy snowfall, even blizzard in the place where cold air meets with warm airs. Like Lesser Snow, Rain Water and Grain Rain, Greater Snow is the solar term which usually reflects precipitation directly.

我国古代将大雪后的十五天分为三候：一候鹖鴠不鸣；二候虎始交；三候荔挺出。意思是说，此时因天气寒冷，寒号鸟也不再鸣叫了；此时是阴气最盛的时期，正所谓盛极而衰，阳气已有所萌动，所以老虎开始有求偶行为；荔挺为兰草的一种，也开始抽发新芽。但是，基于地理位置的原因，包括资阳地区在内的川西平原，却常常是"大雪"时节不见雪，更多时候是伴以阴冷少雨天气，年平均最低气温也多在5℃以上。

Ancient China has divided the 15 days after Greater Snow into 3 pentads. In the first pentad, Hanhao bird stops singing because of cold weather. In the second pentad, yin qi is the strongest and yang qi begins to grow. Tiger starts to look for mate. In the third pentad, Li Ting, a kind of orchid, begins to sprout. However, it is hardly snow in the Western Sichuan Plain including Ziyang region in Greater Snow because of geographical location. Normally, it is gloomy and less raining, and the average temperature is above 5℃.

食材生产 | PRODUCING FOOD INGREDIENTS

大雪时节，气候越发寒冷，中国北方大部分地区已是千里冰封、万里雪飘，但在四川盆地，乡野间仍然充满绿色的生机，不仅冬小麦长势喜人，而且地里还有芹菜、萝卜、青菜、白菜、莲花白、菠菜等新鲜蔬菜和草莓、柚子等水果。

It is getting colder in Greater Snow. Most parts of Northern China has been frozen. However, the rural area of Sichuan Basin is full of green vitality. Winter wheat is growing well, and people can find celery, radish, pakchoi, Chinese cabbage, cabbage, spinach, strawberry, pomelo in the fields.

饮食养生 | DIETARY REGIMEN

大雪后天气寒冷，阳气潜伏、阴气旺盛，故不仅要坚持冬季养肾的总体养生原则，还要格外“大补”热量。因为肾脏“恶寒”，而此节气十分寒冷，因此，人体祛寒益肾、养阴生精显得更加重要，可适量进食一些高热量、高蛋白、高脂肪的食物，以抵御寒气，增强机体的抗寒能力。我国传统医学和养生学认为，大雪是“进补”的最佳时节，但不可太过或乱补，以免影响健康。此节气宜温补，在调味品中可适当使用花椒、肉桂、茴香等温热调味料，以助阳气升发，促进血液循环，从而达到养肾祛寒、助阳强身的作用。此时的大白菜清甜细嫩，对有内热的人而言，尤其适宜。

It is very cold after Greater Snow. Yang qi has lurked and yin qi has strengthen. Therefore, people should not only follow the principle of nourishing kidney in winter, but also get extra heat energy. Human kidney hates coldness, hence it is more important to tonify kidney, dispel coldness, nourish yin and engender essence in this cold solar term. People can take some food with high caloric value, high protein and high fat which can prevent coldness and enhance body's cold resistance. TCM and regimen believe that Greater Snow is the best time to tonify the body. However, we should tonify the body in a proper way, otherwise our health would be affected. In order to tonify kidney, dispel coldness, enhance yang and strengthen the body, we'd better use warm tonic spices like Sichuan pepper, cinnamon, fennel, which can help yang qi grow and promote blood circulation. Chinese cabbage in this time is sweet and delicate, which is good for those people who have too much internal heat.

大雪美食 | FINE FOODS IN GREATER SNOW

大雪的主要食俗，一是“喝粥养生”，以利温胃保暖；二是腌菜腌肉。民谚说“小雪腌菜，大雪腌肉”，“未曾过年，先肥屋檐”，形象地描绘了大雪节气的饮食风俗。大雪节气一到，家家户户都忙着腌制“咸货”，无论是家禽还是鱼肉，人们用传统方法加工成香气袭人的美食，以迎接即将到来的新年。而在四川地区，最符合大雪节气的美食则是红彤彤、热辣辣的“川式火锅”，亲朋好友围炉而坐，欢聚一堂，尽扫冬日阴霾。

As for the main eating customs in Greater Snow, the fist one is "eating porridge to preserve the health". Porridge can warm the stomach and keep the body warm. The second one is making pickles and preserved meat. As the proverbs go, "preserve vegetables in Lesser Snow and preserve meat in Greater Snow" and "hang Chinese bacon from the eaves before Spring Festival". Such proverbs have described the eating customs in Greater Snow vividly. When Greater Snow arrives, every family will make preserved foods, no matter poultry or fish. People make delicious foods in traditional way to welcome the coming new year. In Sichuan region, the best food for Greater Snow is red and spicy "Sichuan Hot Pot". Relatives and friends sit around the pot, enjoy the happy time and dispel the gloom of winter day.

鸳鸯火锅

Double-Flavor Hot Pot

Gala

鸳鸯火锅

Double-Flavor Hot Pot

食材配方

1.红锅卤汁原料：姜片80克　大蒜30克　葱段30克　干辣椒节100克　干花椒30克　糍粑辣椒500克　郫县豆瓣175克　豆豉10克　醪糟25克　草果5克　八角5克　山柰5克　桂皮5克　丁香2克　灵草3克　排草2克　白豆蔻3克　小茴香3克　香叶2克　冰糖50克　胡椒粉2克　鸡精5克　味精2克　料酒10毫升　棒骨汤2000毫升　熟菜籽油700毫升　牛油500克

2.白锅卤汁原料：姜片5克　葱节10克　白胡椒粉2克　红枣4～5颗　枸杞10～15粒　番茄片3片　食盐5克　鲫鱼2条（约200克）　鸡精5克　味精2克　鸡油35克

3.味碟原料：蒜泥50克　芝麻油250毫升　食盐20克　味精20克

4.涮锅食材：鲜毛肚150克　鲜黄喉150克　鲜鹅肠150克　鲜鳝鱼片150克　腰片150克　牛肉片150克　肉丸子150克　竹笋100克　苕粉100克　豆腐皮100克　香菇100克　冬瓜100克　海带100克　平菇100克　金针菇100克　大白菜100克　藕片100克　土豆片100克

制作工艺

①锅置火上，加入熟菜籽油、牛油加热至120℃时，放入干辣椒、干花椒炒香后捞出待用，再放入郫县豆瓣酱、糍粑辣椒、姜片、大蒜、葱段、豆豉、草果、八角、山柰、桂皮、丁香、灵草、排草、白豆蔻、小茴香、香叶炒香，入棒骨汤、料酒、醪糟、冰糖、胡椒粉、鸡精、味精烧沸出味，制成红锅卤汁。

②锅置火上，加入棒骨汤、姜片、葱节、白胡椒粉、食盐、红枣、枸杞、番茄片、鲫鱼、鸡精、味精、鸡油烧沸，制成白锅卤汁。

③取一鸳鸯火锅，在“S”形的一格放入红锅卤汁，另一格放入白锅卤汁，配上味碟原料和涮锅食材上桌即成。

风味特色

此火锅红味色泽红亮、麻辣鲜香；白味咸鲜香浓，食材丰富，质地多样。

创意设计

唐代白居易《问刘十九》诗言：“绿蚁新醅酒，红泥小火炉。晚来天欲雪，能饮一杯无？”四川火锅传承了该诗所描绘的有声有色、有形有态、有情有意的精髓。在鸳鸯火锅中，红锅卤汁虽然加入了大量花椒、辣椒及辛燥香料，但有菜籽油的加入，能解毒润燥；白锅卤汁则清鲜平和，而且吃火锅时加入豆腐及时蔬烫煮，更能缓和热性及燥辣之味。大雪时节，围炉而坐，不仅温补，更暖人心。

Ingredients

1. Spicy Soup: 80g ginger, sliced; 30g garlic; 30g spring onions, segmented; 100g dried chilies, segmented; 30g dried Sichuan peppercorns; 500g Guizhou Ciba chilies; 175g Pixian chili bean paste; 10g fermented soybeans; 25g fermented glutinous rice; 5g black cardamom; 5g star anise; 5g sand ginger; 5g cinnamon; 2g cloves; 3g Chinese liquorice; 2g nephrolepis; 3g round cardamom (Amomum kravanh); 3g fennel; 2g bay leaves; 50g rock candy; 2g pepper; 5g chicken essence; 2g MSG; 10ml Shaoxing cooking wine; 2,000ml pork leg bone stock ; 700ml precooked rapeseed oil; 500g beef tallow

2. Milky Soup: 5g ginger, sliced; 10g spring onions, segmented; 2g ground white pepper; 4-5 red dates; 10-15 wolfberries; 3 tomato slices; 5g salt; 2 crucian carps (about 200g); 5g chicken essence; 2g MSG; 35g chicken fat

3. Dipping Sauce: 50g garlic, crushed; 250ml sesame oil; 20g salt; 20g MSG

4. Ingredients to boil: 150g water-soaked pork tripe; 150g cow throat; 150g goose intestines; 150g eels, sliced; 150g pork kidneys, sliced; 150g beef, sliced; 150g meatballs; 100g bamboo shoots; 100g sweet potato noodles; 100g sheets of bean curd; 100g shitake mushrooms; 100g winter melon; 100g kelp; 100g needle mushroom; 100g oyster mushrooms; 100g Chinese cabbage; 100g lotus root, sliced; 100g potatoes, sliced

Preparation

1. Combine the precooked rapeseed oil and beef tallow in a wok and heat up to 120℃. Add dried chilies and dried Sichuan peppercorns, and stir fry till aromatic and remove. Add Pixian chili bean paste, Guizhou Ciba chilies, ginger slices, garlic, spring onion segments, fermented soybeans, black cardamom, star anise, sand ginger, cinnamon, cloves, liquorice, nephrolepis, round cardamom, fennel and bay leaves, and stir fry till aromatic. Add stock, Shaoxing cooking wine, fermented glutinous rice, rock candy, pepper, chicken essence and MSG, and bring to a boil to make the spicy soup.
2. Add stock, ginger slices, spring onion segments, pepper, salt, red dates, wolfberries, tomato slices, crucian, chicken essence, MSG and chicken fat into a wok and bring to a boil to make the milky soup.
3. Add the two kinds of soup respectively to the two different compartments of a pot partitioned by an "S" -shaped dividing bar. Serve the pot, dipping sauce and ingredients to boil.

Features

aromatic, pungent and numbing spicy soup with fiery and bright colors; delicate and savory milky soup with varied ingredients

Notes

In his poem "Asking Liu Shijiu", the famous Tang Dynasty poet Bai Juyi wrote, "Green Ants a newly brew, red clay miniature stove. With dusk, the snow impending, care to have a cup, or no?" Sichuan hotpot inherits the essence of the vivid and impressive pictures depicted in the poem. Although the spicy soup contains various pungent spices like Sichuan peppercorns and chilies, the rapeseed oil added helps to generate body fluid, clear away heat and detoxify the body. Instead, the milky soup is mild in nature, and people can boil doufu and vegetables in it, which is good for alleviating the pungent taste. In times of Greater Snow, people gather together and eat hotpot around the table, warming both the body and the heart.

白菜鸡肉卷

Chicken Rolls with Cabbages

食材配方

莲花白叶300克　鸡脯肉200克　鸡纵菌50克　胡萝卜50克　姜片5克
鸡蛋清100毫升　姜葱水35毫升　葱段10克　料酒5毫升　胡椒粉1克　鸡精2克
食盐3克　芝麻油3毫升　水淀粉45克　鲜汤150毫升　化猪油30毫升

制作工艺

① 莲花白叶入沸水锅中焯水后捞出晾凉；鸡纵菌焯水后切碎；胡萝卜切碎。

② 将鸡脯肉放入搅拌机中，加入食盐、鸡精、姜葱水、鸡蛋清、胡椒粉、水淀粉、鲜汤搅打成鸡肉蓉，再加入鸡纵菌碎、胡萝卜碎搅匀成馅料。

③ 莲花白叶平铺，放入鸡肉馅料，包卷成条，入笼蒸熟，取出后斜刀切成长约4厘米、厚约2厘米的菱形片，摆放在盘中成花形，点缀上胡萝卜花和鸡枞枞菌。

④ 锅置火上，入油烧至120℃时，放入姜片、葱段爆香，掺入鲜汤烧沸，捞出姜、葱，下食盐、鸡精、芝麻油、水淀粉收汁至浓稠，起锅淋在菜卷上即成。

风味特色

形态美观，鸡肉鲜嫩，味道咸鲜清香。

创意设计

此菜用焯熟后的莲花白叶做皮，包裹鸡肉和鸡枞菌、胡萝卜等原料，口感咸鲜清淡，十分适合老年人和儿童食用。在养生方面，莲花白通利肠胃，鸡肉温中补气、益精填髓，与鸡枞菌、胡萝卜搭配，益胃补血，温补之中不失清润，颇具食疗价值。

Ingredients

300g cabbages; 200g chicken breast; 50g jizong mushrooms; 50g carrots; 5g ginger, sliced; 100ml egg white; 35ml ginger and spring onion juice; 10g spring onions, segmented; 5ml Shaoxing cooking wine; 1g pepper; 2g chicken essence; 3g salt; 3ml sesame oil; 45g average batter; 150ml stock; 30ml lard

Preparation

1. Blanch the cabbages leaves and remove to cool. Blanch the jizong mushrooms and chop finely. Finely chop the carrots.
2. Put the chicken breast in a blender, and add salt, chicken essence, ginger and spring onions juice, egg white, ground pepper, average batter and stock and mince all the ingredients. Add the finely chopped jizong mushrooms and carrots, mix well to make the stuffing.
3. Flatten the cabbage leaves and add the stuffing. Roll up and steam in a steamer till cooked through. Cut the roll by a slant angle into 4cm long and 2cm thick diamond slices. Arrange the slices in a serving plate like a flower, and garnish with carrots and jizong mushrooms.
4. Heat up the cooking oil to 120℃ in a wok, and add ginger slices and spring onion segments and stir fry till aromatic. Add stock, bring to a boil, and remove ginger and spring onions. Add salt, chicken essence, sesame oil and batter, and braise till the soup is thickened. Pour the soup over the chicken slices.

Features

beautiful appearances; tender chicken; savory and delicate tastes

Notes

This dish uses cabbage leaves as wrappers and chicken, jizong mushrooms and carrots as the stuffing. It is savory and delicate in taste, which is suitable for the elders and children. In food therapy, cabbages are good for moistening the intestines and nourishing the stomach. Chicken meat is mild in nature and helps to nourish qi. With jizong mushrooms and carrots, the dish has functions of nourishing the blood and tonifying the stomach.

蜜汁红薯

Fried Sweet Potatoes with Syrup

食材配方

安岳红薯200克　糖浆30克　食用油1000毫升（约耗50毫升）

制作工艺

①红薯去皮，切成长约6厘米，横截面边宽皆为2厘米的长条。

②锅置火上，入油烧热，将红薯条入锅，炸至金黄色时捞出，摆入盘中，配以糖浆即成。

风味特色

薯条色泽金黄，质地软糯，味道甜香。

创意设计

安岳红薯产量大、品质优，是资阳著名的地方特产。大寒前后，四川许多地区的人们都特别喜食红薯，烤红薯成为一道独特的美食风景。此款蜜汁红薯经煎炸后蘸取糖浆食用，是对红薯的一种创新做法。在养生方面，红薯补中和血、益气宽肠，富含膳食纤维、B族维生素，能促进胃肠蠕动、胆固醇排泄及防止动脉硬化，搭配糖浆食用，甘味入脾，更能强健脾胃，可为冬至进一步滋补打下基础。

Ingredients

200g Anyue sweet potatoes; 30g syrup; 1,000ml cooking oil (about 50ml to be consumed)

Preparation

1. Peel the sweet potatoes and cut into 6cm long, 2cm wide and 2cm thick strips.
2. Heat up the cooking oil in a wok, and deep fry the sweet potato strips till golden brown. Transfer to a serving dish, and serve with the syrup.

Features

golden brown color; tender, glutinous and sweet tastes

Notes

Anyue sweet potato is a famous Ziyang specialty with high quality. Around the time of Greater Snow, people prefer to eat sweet potatoes in many places of Sichuan Province. Roasting sweet potato becomes a unique culture of gourmet. This innovative dish is a deep fry of sweet potatoes served with syrup. In food therapy, sweet potatoes help to nourish the blood, moisten the intestines and invigorate qi. It is rich in dietary fiber and B-group vitamin, which accelerates the gastrointestinal motility and cholesterol excretion and prevents the arteriosclerosis. The sweet syrup helps to tonify the spleen and nourish the stomach. It helps to lay a solid foundation for taking tonics during the upcoming the Winter Solstice.

THE WINTER SOLSTICE

冬至
日照数九清霜风高
冬至日思亲
明·于谦
客里逢佳节，天涯忆老亲。
葭灰初应候，梅蕊渐回春。
醉讶朱颜好，愁添白发新。
孤云恒在望，翘首欲沾巾。

冬至物候 | PHENOLOGY IN THE WINTER SOLSTICE

冬至是二十四节气中的第二十二个节气，也是我国古人最为看重的节气之一。当太阳到达黄经270° 时为冬至，时间为每年的12月21～23日。冬至这一天是北半球全年中白天最短、夜晚最长的一天。《二十四节气集解》云："阴极而至，阳气始生，日南至，日短之至，日影长至，故曰冬至。"冬至又称为"冬节""至节""长日"，过了冬至，白天变长而夜晚缩短，有"冬至一阳生"之说，也同时意味着天气进入一年中最寒冷的季节。

The Winter Solstice is the twenty-second of the 24 Solar Terms. It is one the most important solar term for ancient Chinese. It begins when the sun reaches the celestial longitude of 270°. It usually begins from 21st to 23rd December every year. On the day of the Winter Solstice, the day time is the shortest and the night time is the longest in a year in the Northern Hemisphere. The book the *Collective Interpretation of the Twenty-Four Solar Terms* said, "Yin is at the peak. Yang qi starts to grow. The sun is at the extremely south. Day time is the shortest. Shadow is the longest. Hence, it is called the Winter Solstice". The Winter Solstice is called Winter Festival, Solstice Festival and Long Day. After the Winter Solstice, day time will become longer and night time will become shorter. Since the day of the Winter Solstice, yang qi has begun to grow. After that day, people will experience the coldest season in a year.

我国古代将冬至后的十五天分为三候:一候蚯蚓结；二候麋角解；三候水泉动。古人认为，蚯蚓是阴曲阳伸之物，此时阳气虽已萌生，但阴气强盛，土中的蚯蚓仍然蜷缩着身体。古人还认为，麋与鹿同科，却阴阳不同，鹿角朝前生，所以属阳，故而夏至时"鹿角解"；而麋的角朝后生，所以属阴，此时阳气始生，而阴气开始减弱，所以阴性的麋角到冬至后便慢慢开始脱落。山中的泉水此时也逐渐开始流动。

Ancient China has divided the 15 days after the Winter Solstice into 3 pentads. In the first pentad, earthworm curls up underground. Ancient Chinese believed that earthworm will curl up in yin situation and will strengthen in yang situation. In this period, although yang qi has started to grow, yin qi is still very strong, hence earthworm still curl up in the earth. In the second pentad, elk horn has fallen. Ancient Chinese believed that deer horn which belongs to yang curves forward, and it will fall in the Summer Solstice. Elk horn which belongs to yin curves backward, and it will fall in the Winter Solstice. In the third pentad, fountain in mountains begins to flow.

食材生产 | PRODUCING FOOD INGREDIENTS

冬至时节，气温持续降低，此时正值四川地区小春作物生长发育和田间管理的关键时期，农民依然忙于农活，为第二年的春耕、春播做好准备。包括资阳在内的川西平原，白菜、莲花白、白萝卜等时令蔬菜收获在望。新鲜冬笋是四川人喜食善烹的季节性食材，此时也是四川各地冬笋上市的大好时节。

In the Winter Solstice, it is getting colder. It is the critical time for the growth of crops sworn in late autumn and the filed management. Farmers are busy with the preparation works for next year's spring ploughing and spring sowing. In the Western Sichuan Plain including Ziyang, seasonal vegetables like Chinese cabbage, cabbage, white radish are going to be picked. Fresh winter bamboo shoots are the favorite seasonal food ingredients for Sichuan people. It is the best time to buy the winter bamboo shoots in the market.

饮食养生 | DIETARY REGIMEN

中国传统医学和养生学认为，人的生命活动和自然环境息息相关。“冬至一阳生”，冬至虽然阴气盛极但潜藏的阳气已开始初生，这是该节气的重要表现。人体中的阳气有推动、温煦、防御、固摄的作用，是构成人体和维持人体生命活动的最基本物质。所以冬至时节，阳气开始生发，需要顺应生发之势、顾护阳气，在饮食养生上，除养阴潜阳、温补阳气外，还需养肾填精、滋养气血，增加高热能食品以增强御寒能力。而牛羊肉属于温补之物，能补益气血，海参色黑入肾、补肾填精，鳝鱼养血润燥，可用牛羊肉、海参等制作菜肴食用。

TCM and regimen believe that human life activity is connected with natural environment directly. “Since the Winter Solstice, yang has grown”. Although yin qi is extremely strong in the Winter Solstice, torpid yang qi starts to grow. It is the important characteristic of the Winter Solstice. Yang qi in human body has the function of promoting, warming, protecting and solidifying. It is the base to form the human body and maintain the life activity. Hence, when yang qi starts to grow in the Winter Solstice, people should follow the trend to protect yang qi. As for eating regimen, people should tonify kidney, nourish qi and blood, and take high calorie food to enhance cold-resistance, in addition to nourish yin, develop and tonify yang. Beef and mutton are warm tonic foods, which can enrich blood. Sea cucumber is back color and can tonify kidney, replenish essence. Eel can enrich blood and moisten dryness. People can make dishes with beef, mutton and sea cucumber.

冬至美食 | FINE FOODS IN THE WINTER SOLSTICE

我国古代人民历来都极为重视冬至这一节气，认为它可与新年媲美，号称“亚岁”或“小岁”。在汉代时期，冬至又称“冬节”，官府要举行 “贺冬”仪式，民间流行“拜冬”之礼。据《后汉书》记载：“冬至前后，君子安身静体，百官绝事，不听政，择吉辰而后省事。”所以这一天官方例行放假，商旅停业，举国休息。冬至祭天盛行在唐朝，至明清时更被推崇，后逐渐流传到民间，演变成冬至祭祖的习俗。在古代，无论是官方祭天，还是民间祭祖，都是期望后来的日子风调雨顺，丰衣足食。在民间，由于人们认为“冬至大如年”，所以，在全国各地均有许多由来已久的饮食习俗及美食品种：在北方地区，吃饺子，有“冬至饺子，夏至面”之语；在南方许多地区，此时则吃馄饨，有“冬馄饨，年馎饦”之语（馄饨，四川称作“抄手”）。相传在阳气始生的冬至日食用馄饨，能够破除阴阳包裹的混沌状态，进而资助阳气的生长。

Ancient Chinese paid great attention to the Winter Solstice. People think it is as important as new year, and call it “Ya Sui” or “Xiao Sui”. In Han Dynasty, the Winter Solstice was called Winter Festival. Government would hold “He Dong” ceremony, and common people would use “Bai Dong” protocol. *Book of Late Han* recorded, “around the Winter Solstice, gentlemen would retreat to a quiet place. Officials would take a break and restart work in an auspicious time”. On the day of the Winter Solstice, government would

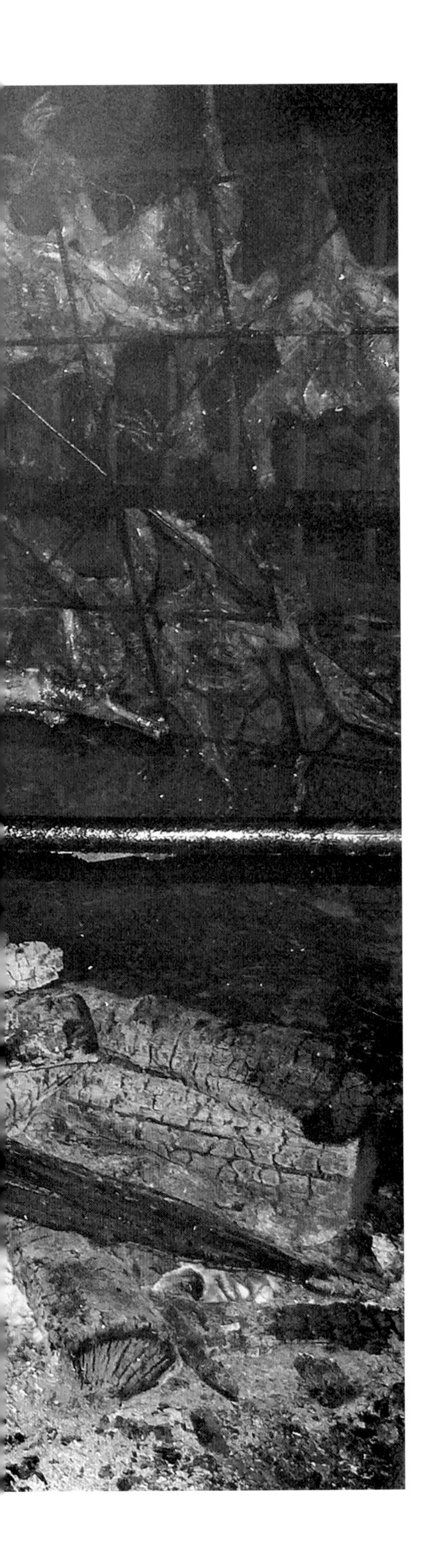

take a holiday, business would stop, and the whole country would take a rest. Worshiping heaven in the Winter Solstice was popular in Tang Dynasty (618-907), and be given more attention to in Ming Dynasty and Qing Dynasty. Later, common people has changed it into a worshiping ancestors custom in the Winter Solstice. In Ancient time, the purpose of worshiping heaven and ancestors is praying for good weather for crops and good harvest in future. Since common people think the Winter Solstice is as important as the new year, there are lots of eating customs and fine foods with long history all over the country. In the North, people eat dumpling. As the proverb goes, "eat dumpling in the Winter Solstice and noodles in the Summer Solstice". In the South, people eat wonton. As the proverb goes, "eat wonton in winter and hakutaku in new year". Wonton is called Chaoshou in Sichuan. According to legend, eating wonton on the day of the Winter Solstice when yang qi starts to grow, will break through the chaos and help the growth of yang qi.

在四川地区，冬至当日还常常杀年猪，用以做“冬至肉”“吃白肉”。《成都通览》言：“冬至日祭祖，杀猪腌过年肉，或装香肠。”白肉是清朝满族人在冬至日祭祖后分食的祭祀用肉，该风俗在清末传入四川，人们在祭祖时使用“刀头肉”，待祭祖仪式结束后，将其切为薄片，调上蒜泥味食用，则成为川菜名品“蒜泥白肉”。时至今日，四川人盛行在冬至时吃羊肉，此时，在资阳、简阳、成都等地的大街小巷都可见“羊肉汤”的身影，人们以此暖和身体、助阳养生。

In Sichuan region, there is a custom of slaughtering pigs for new year on the day of the Winter Solstice. People make the Winter Solstice Meat and eat Bai Rou. *Overall View of Chengdu* has said, "on the day of the Winter Solstice, people worship ancestors, slaughter pigs to make Chinese bacon and sausage." Bai Rou is the meat offered to ancestors by Manchus on the day of the Winter Solstice. This custom has been introduced to Sichuan in Qing Dynasty. After the ceremony, people will slice the meat, and eat with garlic sauce. It has become the famous Sichuan dish "Boiled Pork in Garlic Sauce". Nowadays, Sichuan people usually eat mutton in the Winter Solstice. You will see mutton soup everywhere in Ziyang, Jianyang and Chengdu. It can warm up the body, promote yang and preserve health.

羊肉汤锅

Lamb Soup Hot Pot

食材配方

熟羊肉500克　熟羊杂500克　萝卜片100克　白菜100克　豌豆苗100克　青笋尖100克　食盐20克　胡椒粉2克　料酒20毫升　姜片50克　葱段60克　味精10克　枸杞10克　红枣10克　葱花60克　香菜碎60克　红尖椒碎40克　豆腐乳60克　羊骨汤2000毫升

制作工艺

① 熟羊肉、熟羊杂切成片后装盘备用。

② 葱花、香菜碎、红尖椒碎、豆腐乳、食盐、味精入碗调成蘸碟。

③ 汤锅置火上，加入羊骨汤，下食盐、胡椒粉、姜片、葱段、味精、料酒、枸杞、红枣调成汤汁上桌，配上熟羊肉片、熟羊杂片和时令蔬菜，烫涮食材后蘸味碟食用。

风味特色

汤色乳白，咸鲜香浓，羊肉细嫩，肥而不腻。

创意设计

四川民间有“冬至吃羊肉，暖和一冬天”之语。冬至时，四川许多地方的大街小巷都可见吃羊肉汤的人。此菜以资阳、简阳等地出产的黑山羊为主要食材制成，地方风味浓郁。在养生方面，羊肉味甘、性温，具有补肾壮阳、温补气血、开胃健脾、暖中祛寒等作用，既能抵御风寒、滋补身体，更能弥补四川盆地阳光不足的缺陷，温补之中更显滋润，有利于温阳滋阴。羊肉、羊杂富含人体所需的蛋白质、B族维生素及矿物质，补益肾阳之效显著。

Ingredients

500g precooked lamb meat; 500g precooked lamb offal; 100g white radish, sliced; 100g Chinese Cabbage; 100g pea vine sprouts; 100g stem lettuce tips; 20g salt; 2g pepper; 20ml Shaoxing cooking wine; 50g ginger, sliced; 60g spring onions, segmented; 10g MSG; 10g wolfberries; 10g red dates; 60g spring onions, finely chopped; 60g corianders, finely chopped; 40g red chilies, finely chopped; 60g fermented doufu; 2,000ml lamb bone stock

Preparation

1. Slice the precooked lamb and lamb offal.
2. Combine the finely chopped spring onions, corianders, red chilies, fermented doufu, salt and MSG in a bowl to make dipping sauce.
3. Heat the stock in a pot and add salt, pepper, sliced ginger, segmented spring onions, MSG, Shaoxing cooking wine, wolfberries and red dates to make milky soup. Boil the cooked lamb slices, offal slices and vegetables, and dip into the dipping sauce before eating.

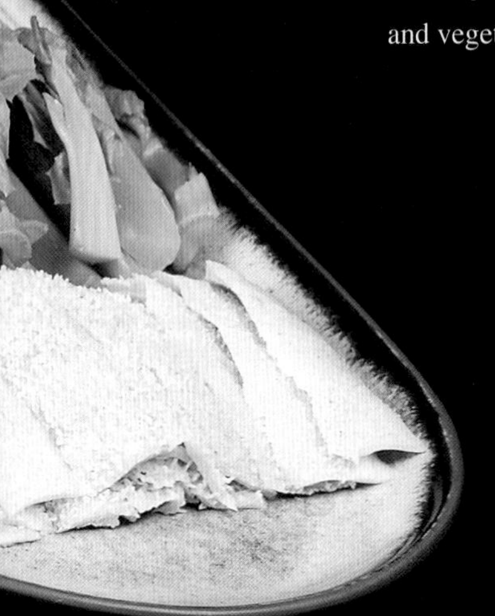

Features

milky soup; savory and aromatic tastes; tender, fatty but not greasy meat

Notes

There is a folk belief in Sichuan that if people eat lamb meat on the the Winter Solstice, they will feel warm the whole winter. On the day of the Winter Solstice, you can see people enjoying lamb soup hot pot in high streets and back lanes. The main ingredient of the hot pot is the black goat meat produced in Ziyang and Jianyang. In food therapy, lamb meat is mild in nature and has functions of invigorating the kidney, nourishing the blood, stimulating the appetite, tonifying the spleen and dispelling cold. It is suitable for people in Sichuan Basin where it is rare to see the Sun. Lamb meat and offal are rich in protein, B-group vitamins and mineral substances, ideal for nourishing the kidney and reinforcing yang.

蒜泥白肉卷

Boiled Pork Rolls in Garlic Sauce

食材配方

带皮猪坐臀肉250克　蒜泥30克

姜片3克　葱段5克　黄瓜100克

食盐3.5克　白糖7克　味精1克

料酒3毫升　复制红酱油40毫升

芝麻油3毫升　红油30毫升

制作工艺

① 锅置火上，注入清水，放入猪肉、姜片、葱段、料酒，煮至猪肉断生后捞出，用原汤浸泡至凉透；黄瓜洗净，片成薄片。

② 将晾凉的猪肉片成长约10厘米、宽约4厘米、厚约0.1厘米的薄片，在肉片上放入黄瓜片，包卷成白肉卷摆放在盘中。

③ 蒜泥、食盐、味精、白糖、复制红酱油、芝麻油、红油入碗调制成蒜泥味汁，最后浇淋在白肉卷上成菜。

风味特色

成菜色泽红亮，肉质软嫩，肥而不腻，味道香辣回甜，蒜香浓郁。

创意设计

冬至时节杀年猪、吃白肉，是清末民国时期四川一些地方的传统习俗。此菜是在川菜传统名品“蒜泥白肉”的基础上改进而成。在养生方面，猪肉补肾滋阴、益气养血；大蒜温中解毒、抑菌防腐，用猪肉搭配极具四川特色的蒜泥味型，不仅口味独特，还能解油腻、助消化、降血脂、抗动脉粥样硬化。

Ingredients

250g pork rump; 30g garlic, crushed; 3g ginger, sliced; 5g spring onions, segmented; 100g cucumber; 3.5g salt; 7g sugar; 1g MSG; 3ml cooking oil; 40ml compound soy sauce; 3ml sesame oil; 30ml chili oil

Preparation

1. Heat water in a wok, add pork rump, ginger, spring onions and cooking oil, and boil till almost cooked. Ladle out the pork and soak in the stock till cold. Rinse the cucumber and slice.
2. Slice the pork 10 cm long, 4 cm wide and 0.1 cm thick. Place cucumber slices on pork slices, roll up the pork slices, and transfer to a serving dish.
3. Mix the garlic, salt, MSG, sugar, compound soy sauce, sesame oil and chili oil to make seasoning, and pour over the rolls.

Features

reddish brown in color; soft and tender in texture; fatty but not greasy; spicy and sweet tastes; an aroma of garlic

Notes

On the the Winter Solstice, it was a custom to slaughter the pig and eat boiled pork in some places of Sichuan back in the period of the Republic of China. Boiled Pork Rolls in Garlic Sauce is an improved version of the traditional dish—Boiled Pork in Garlic Sauce. The improved dish is better for people's health. Pork not only tonifies the kidneys and nourishes yin, but also tonifies qi and nourishes blood. Garlic not only warms the spleen and the stomach, but also prevents bacteria and corrosion. The combination of pork and garlic has an attractive flavor, and it relieves greasiness, reduces blood pressure and prevents atherosclerosis.

滋补抄手

Nourishing Chaoshou

食材配方

抄手皮30张　猪绞肉200克　土鸡一只（约1500克）　当归20克　老姜20克　炙黄芪100克　大葱20克　姜葱水15毫升　炖鸡汤1200毫升　胡椒粉2克　芝麻油5毫升　味精5克　食盐20克

制作工艺

① 土鸡制净，焯水后放入大砂锅中，掺入清水，放入当归、炙黄芪，用小火炖制2个小时以上，制成鸡汤后晾凉备用。

② 猪绞肉中加食盐、胡椒粉、味精搅匀，加入姜葱水搅打至肉质黏稠起胶，再分次加入冷鸡汤搅打至肉质松散，最后加入芝麻油拌匀成馅料。

③ 取抄手皮一张，放入馅料，先对折成三角形，再将左右两角尖向中折叠粘合(粘合处抹少许馅糊)成菱角形制成抄手生坯。

④ 将食盐、胡椒、味精入碗，加入适量原汤。

⑤ 锅置火上，加水烧沸，下抄手生坯，煮熟后捞出放入碗中即成。

风味特色

皮薄爽滑，馅心细嫩多汁，汤浓味美。

创意设计

自宋代以来，我国民间已有在冬至之日吃馄饨的食俗，据宋代陈元靓的《岁时广记》称："京师人家，冬至多食馄饨，故有冬至馄饨年之萌始之说。""抄手"是四川方言对馄饨的别称。在四川各地都有特色不一的抄手，深受百姓喜爱。此款滋补抄手，是将养生滋补的当归炖鸡与抄手完美融合的结果。在养生方面，当归、黄芪均为药食两用之物，能补血活血、补气健脾，所含活性成分能延缓人体衰老，并有抗氧化、抗肿瘤等作用，再辅之以猪肉、鸡肉及温中补气的面粉，气血双补作用更为突出。

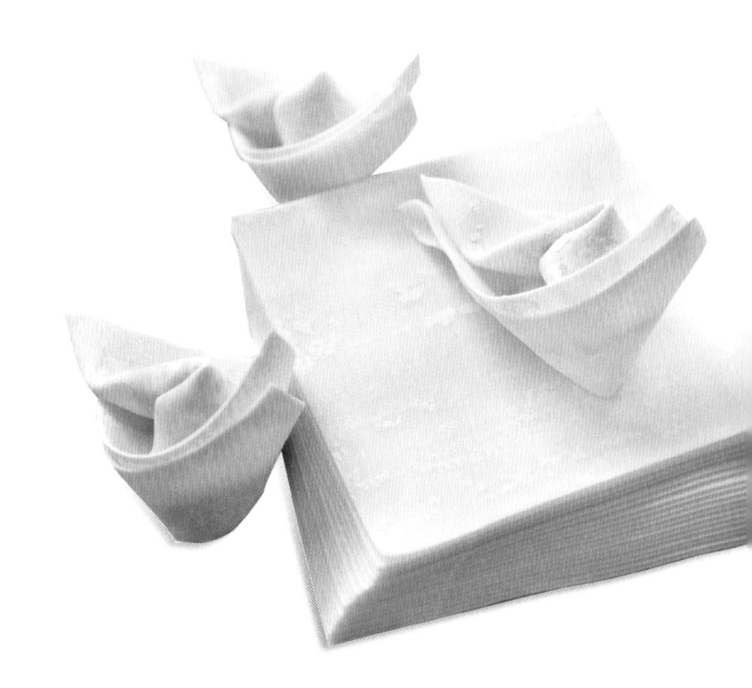

Ingredients

30 pieces of chaoshou wrappers; 200g ground pork; 1 free-range chicken (about 1,500g) ; 20g danggui (Angelica sinensis); 20g ginger; 100g Mongolian milkvetch; 20g spring onions; 15ml garlic and spring onion juice; 1,200ml chicken stock; 2g pepper; 30g egg; 5ml sesame oil; 5g MSG; 20g salt

Preparation

1. Blanch the free-range chicken and remove to a big casserole. Add water, danggui and Mongolian milkvetch, and simmer over a low flame for over 2 hours. Wait until the chicken soup is cool.
2. Rub the ground pork with salt, pepper and MSG. Add ginger and spring onion juice and beat until thick and glutinous. Add the cold chicken soup and beat until the pork becomes loose. Mix with sesame oil for stuffing.
3. Put the stuffing on the wrapper. Fold up in the shape of triangle, and stick the left and right corners together with stuffing juice to make Chaoshou.
4. Combine salt, pepper and MSG in a bowl, and add the chicken soup.
5. Heat water in a wok till boiling. Boil the Chaoshou till cooked through, and transfer to the bowl.

Features

smooth thin Chaoshou skin; tender and juicy stuffing; delicious soup; high nutritional value

Notes

People in China eat wontons on the Winter Solstice, a tradition dating back to the Song Dynasty. Chaoshou is Sichuan version of wontons. In different places of Sichuan, people make different chaoshou. This dish is a great nutritious combination of stewed chicken with chaoshou. Danggui and Mongolian milkvetch have health care values of enriching the blood, promoting blood circulation, tonifying qi and strengthening the spleen. Active constituents in these two ingredients prevent aging, oxidation and cancer. Pork, chicken and flour strengthen the effects of tonifying qi and nourishing the blood.

咏廿四气诗·小寒十二月节
唐·元稹
小寒连大吕，欢鹊垒新巢。
拾食寻河曲，衔紫绕树梢。
霜鹰近北首，雊雉隐丛茅。
莫怪严凝切，春冬正月交。
LESSER COLD

小寒
百草减翠
疏木摇空

小寒物候 | PHENOLOGY IN LESSER COLD

小寒是二十四节气中的第二十三个节气。当太阳到达黄经285° 时为小寒，时间为每年的1月5日～7日。元朝吴澄《月令七十二候集解》言："小寒，十二月节。月初寒尚小，故云，月半则大矣。"此节气是全年中仅次于大寒的极寒冷节气。

Lesser Cold is the twenty-third of the 24 Solar Terms. It begins when the sun reaches the celestial longitude of 285°. It usually begins from 5th to 7th January every year. In his book *A Collective Interpretation of the Seventy-Two Phenological Terms*, Wu Cheng (Yuan Dynasty 1271-1368) said, "Lesser Cold is December (lunar calendar) Festival. It is not so cold at the beginning of this month, but it will be severely cold in the second half of this month." The weather in Lesser Cold is the second coldest in a year, next to Greater Cold.

我国古代将小寒后的十五天分为三候：一候雁北乡；二候鹊始巢；三候雉始雊。小寒五日后，虽然天气寒冷，阴气极盛，但是阳气已经生发，大雁首先感受到这个变化，开始从南向北飞；再过五日，喜鹊开始飞回筑巢；又经五日，雉鸡开始啼叫。"雉"即野鸡，雌雄野鸡同时鸣叫谓之"雊"。小寒时节，梅花初开，开始了"花信"。小寒节气相应的花信为：一候梅花；二候山茶；三候水仙。

Ancient China has divided the 15 days after Lesser Cold into 3 pentads. In the first pentad, although it is cold and yin qi is extremely strong, yang qi has started to grow. Wild goose has felt the change and begun to fly from the south to the north. In the second pentad, magpie has come back to nest. In the third pentad, pheasant starts to sing. Pheasant is called "Zhi" in China. When male pheasant and female pheasant sing together, Chinese people call it "Gou". In Lesser Cold, wintersweet begins to blossom. The news of flowers blooming in Lesser Cold are as follows: wintersweet flower for the first pentad, camellia for the second pentad and narcissus for the third pentad.

食材生产 | PRODUCING FOOD INGREDIENTS

民谚语："小寒大寒，冷成冰团。"小寒时节的农业生产以加强越冬作物管理为主。而在四川地区，小寒时节常常雾重霜凝，对来年春耕春种极有好处，正如民谚所言："小寒节日雾，来年五谷福。"此时，豌豆尖、冬寒菜等当季蔬菜陆续上市，成为川人隆冬季节餐桌上少不了的一抹翠绿。

As the proverb goes, "Be frozen into ice ball in Lesser Cold and Greater Cold." The main job in fields is to strengthen the management of overwintering crops in Lesser Cold. In Sichuan region, it is normally heavily foggy and frosty in Lesser Cold, which is good for the spring ploughing and spring sowing in next year. As the saying goes, "If it is foggy in Lesser Cold, there will be good harvest in the coming year". Seasonable vegetables like snow pea shoots, Chinese mallow are on the market at this moment. They are indispensable to Sichuan people's dinning-table in midwinter.

饮食养生 | DIETARY REGIMEN

小寒时节天气极为寒冷，虽然阳气已动，但阴气极盛，人体易气血凝滞，需注意补肾护阳，在饮食上应合理进补，及时补充气血津液，以便抵御严寒侵袭，强壮身体。进补时应以温补为宜，可选用白萝卜、白菜、羊肉、猪肉、鸡肉、鸽子、鳝鱼、虾、核桃、大枣、桂圆、芝麻、山药、莲子、百合、栗子等，菜肴可选择山药炖羊肉、大蒜烧鳝鱼等。此外，《千金方》还针对一些寒性疾病，提出了“冬服药酒两三剂，立春则止，终身常乐，百病不生”的养生理念，因此，可根据个人身体状况在此时适当选服药酒，但必须注意限量饮用。

It is very cold in Lesser Cold. Although yang qi has started to grow, yin qi is extremely strong and human body will suffer from the stagnation of qi and blood. Hence, people should tonify kidney and protect yang. In terms of diet, we should tonify qi, enrich blood and nourish body fluid to keep health and prevent coldness. We should take warm tonifying foods, like white radish, Chinese cabbage, mutton, pork, chicken, pigeon, eel, shrimp, walnut, red date, longan, sesame, Chinese yam, lotus seed, lily, chestnut, etc. We can cook mutton with Chinese yam, and cook eel with garlic. Moreover, aiming at some cold diseases, *Thousand Golden Prescriptions* has pointed out following health concept. “Taking two or three doses of herbal liquor in winter before the Beginning of Summer can prevent all kinds of diseases”. Therefore, we can take proper and limited amount of herbal liquor in this period according to individual health conditions.

小寒美食 | FINE FOODS IN LESSER COLD

小寒时节，在许多年份已进入农历十二月，即腊月。从先秦开始，“腊祭”是向百神报告“年丰物阜”，感谢百神保佑、使得农业大获丰收的仪式。此外，还要举行傩戏表演或驱鬼驱疫活动。“小寒忙买办，大寒忙过年”，人们开始置办年货，到集市购买红纸、年画、鞭炮、糖果、彩灯等，为接下来的春节做准备。到宋朝时，受佛教影响，祭祀众神的腊日慢慢演变为庆贺佛祖诞辰的浴佛日，并逐渐成为腊八节。小寒节气的时间，大多年份与农历十二月初八的腊八节相近。在腊八节时，最重要的美食就是“腊八粥”。据宋朝孟元老《东京梦华录》载：“诸大寺作浴佛会，并送七宝五味粥与门徒，谓之腊八粥。”元、明、清以后，腊八粥更是风行于大江南北。各地的腊八粥原料、口感略有不同。四川地区的腊八粥除了有果品外，还有蔬菜和肉类。据《华阳县志》载：“十二月八日，俗谓腊八，人家多用各色豆、米、菜、果合煮，间有杂以鸡、鱼、猪肉，谓之腊八粥。”此外，在小寒时节，四川地区的许多家庭还有制作醪糟的习俗，即用糯米发酵的一种风味食品，再与汤圆粉团、鸡蛋等煮制而成醪糟汤圆、醪糟蛋，口味甘甜醇美，略带酒香，最宜严冬食用。

In many years, Lesser Cold is in the twelfth month of the lunar year which is called “La Yue” in Chinese. Since Pre-Qin Times, the purpose of La Worship was to report good harvest to gods. Meanwhile, there were operas and exorcisms. “People are busy with making preparation for the new year in Lesser Cold and with celebrating the Spring Festival in Greater Cold”. People will go to bazaar purchasing red paper, New Year pictures, firecracker, candy, coloured lantern, etc. for the coming Spring Festival. In Song Dynasty, worshiping gods has changed into Bathing Buddha Ceremony for Buddha’s birthday because of the influence of Buddhism, and then gradually changed into Laba Festival. In many years, the date of Lesser Cold is close to the 8th December in lunar year, which is Laba Festival. In Laba Festival, the most important food is Laba Congee. In his book *Records of Dreamily Prosperous Eastern Capital*, Meng Yuanlao has recorded, “All big temples host Bathing Buddha Ceremony, and will share rice porridge with nuts and dry fruits with followers, which is called Laba Congee.” After Yuan Dynasty, Ming Dynasty and Qing Dynasty, Laba Congee has become popular all over the country. The tastes of Laba Congee from different places have tiny difference because of the difference of materials. In Sichuan region, people add vegetables and meat in Laba Congee in addtion to nuts and dried fruits. *Huayang County Local Chronicles* recorded, “the 8th December is calls Laba. People cook rice, different beans, vegetables, fruits together, and add chicken, fish and pork into it, which is called Laba Congee.” Moreover, many Sichuan families make fermented glutinous rice in Lesser Cold, and then cook glue pudding and egg with it. It is sweet, delicious and with some aroma. Server winter is the best time to eat it.

孜香烤羊排

Roasted Lamb Chops with Cumin

食材配方

羊排10块（约1000克） 土豆块100克 姜米10克 蒜米20克
香辣脆椒50克 火锅底料30克 味精2克 白糖2克 食用油70毫升
川味老卤水1000毫升

制作工艺

① 羊排经刀工处理后焯水，入卤水中卤制40分钟后捞出；土豆切块后焯水。
② 烤箱预热至200℃，放入羊排烤制5分钟。
③ 锅置火上，入油烧至160℃时，放入土豆炸至表面色金黄时捞出。
④ 锅置火上，入油烧至100℃时，先放入火锅底料、姜米、蒜米、味精、白糖炒香，再加入香辣脆椒、羊排、土豆块炒香后起锅装盘即成。

风味特色

成菜色泽红亮，外焦里嫩，质地酥软，味道麻辣，香味浓郁。

创意设计

四川资阳地区有一道名菜叫“黑山羊排”，深受当地人喜爱。此菜借鉴法式烤羊排的思路，采用中西嫁接的方式制成，地域特色非常浓烈。在养生方面，冬季食用羊肉具有补血、补阳等功效。川味卤水辛香开胃、不辣不燥，用来卤煮肉类，既能去腥解腻，也适宜冬季温补的需要。

Ingredients

10 lamb chops (about 1,000g); 100g potatoes; 10g ginger, finely chopped; 20g garlic, finely chopped; 50g crispy chilies; 30g hotpot soup base; 2g MSG; 2g sugar; 70ml cooking oil; 1,000ml Sichuan-style spiced broth

Preparation

1. Blanch the lamb chops, remove into the spiced broth, boil for 40 minutes and ladle out. Chop the potatoes, and blanch.
2. Turn on the oven and set the temperature at 200 ℃, and roast the lamb chops for 5 minutes.
3. Heat oil in a wok to 160℃, and deep fry the potato chops till golden.
4. Heat oil in a wok to 100℃. Add the hotpot soup base, ginger, garlic, MSG, sugar, and stir fry till aromatic. Add the crispy chilies, lamb chops and potato chops, stir well and transfer to a serving dish.

Features

reddish brown in color; crispy crusts; spicy taste

Notes

Black Goat Chops is a famous popular dish in Ziyang City. Roasted Lamb Chops with Cumin is made by integrating Chinese and western cooking methods. For instance, the lamb is roasted in a French way but stir fried in a Chinese way. This dish is great for people's health. Having lamb in winter helps nourish the blood and yang. Sichuan-style spiced broth, mild in nature, can whet the appetite and the lamb boiled in the broth have been removed of the odor and grease. This dish is a good choice to nourish the body especially in winter.

红烧狮子头

Braised Meatballs in Brown Sauce

食材配方

去皮猪五花肉500克　荸荠100克　熟火腿25克　冬笋100克　水发香菇50克　菜心50克　金钩10克　鸡蛋2个　老抽5毫升　东古一品鲜10毫升　味精2克　食盐2克　芝麻油3毫升　水淀粉60克　姜葱水200毫升　料酒10毫升　鲜汤300毫升　食用油1000毫升（约耗50毫升）

制作工艺

① 去皮猪五花肉切成约0.3厘米见方的颗粒；荸荠、金钩分别切成绿豆粒大小的颗粒；熟火腿切成长约5厘米、宽约2厘米、厚约0.3厘米的片；冬笋、水发香菇分别切成牙瓣状。

② 猪肉入盆，放入食盐、姜葱水、鸡蛋液、水淀粉搅匀，再入荸荠粒、金钩粒拌匀成肉馅。

③ 锅置火上，入油烧至150℃时，将肉馅做成每个重约60克的肉圆子入锅炸至表面色金黄后捞出放入盆中，加入鲜汤、食盐、料酒、老抽、东古一品鲜，入笼蒸制两小时后出笼，倒入锅中，加入熟火腿、冬笋、水发香菇、菜心烧约20分钟，再入味精、芝麻油、水淀粉收浓汤汁后起锅装盘即成。

风味特色

成菜色泽棕红，质地细嫩，味道咸鲜香浓。

创意设计

红烧狮子头是川菜的一道传统名品。此菜以肥瘦兼备的猪肉与荸荠、香菇、冬笋等为食材制成，更是风味别具。在养生方面，荸荠润燥生津，冬笋化痰消胀，均为时令佳品，搭配猪肉滋阴润燥，实为冬季养阴佳肴。此外，荸荠所含荸荠英能抗菌消炎，冬笋纤维素含量丰富，对降低血压、胆固醇和血脂能起到辅助作用。

Ingredients

500g pork belly; 100g water chestnuts; 25g cooked ham; 100g winter bamboo shoots; 50g water soaked shitake mushrooms; 50g choy sum; 10g dried, shelled shrimps; 2 eggs; 5ml dark soy sauce; 10ml Donggu soy sauce; 2g MSG; 2g salt; 3ml sesame oil; 60g average batter; 200ml garlic and spring onion juice; 10ml Shaoxing Cooking Wine; 300ml stock; 1,000ml cooking oil (about 50ml to be consumed)

Preparation

1. Dice the pork belly into 0.3cm cubic granules. Dice the water chestnuts and the dried, shelled shrimps into mung bean-sized granules. Cut the cooked ham into slices 5cm long, 2cm wide and 0.3cm thick. Cut the winter bamboo shoots and soaked mushrooms in the shape of petals.
2. Combine the pork, salt, ginger and spring onion juice, beaten eggs, batter in a basin, and blend well. Add water chestnuts and dried, shelled shrimps, and mix well.
3. Heat the cooking oil in a wok to 150℃. Make meatballs, 60g each, with the pork mixture, deep fry till golden on the surface, ladle out and transfer to a basin. Add stock, salt, cooking wine, dark soy sauce and Donggu soy sauce to the basin, and steam for two hours. Pour the contents in the basin to a wok, add the ham, winter bamboo shoots, mushrooms and choy sum, and braise for about 20 minutes. Add MSG, sesame oil and batter, continue to braise to reduce and thicken the sauce. Transfer to a serving dish.

Features

reddish brown in color; soft and tender in texture; savory and slightly spicy tastes.

Notes

Braised Meatballs in Brown Sauce is one of the famous traditional Sichuan dishes. It is made from pork (with fat and lean meat), water chestnuts, mushrooms, winter bamboo shoots and other ingredients. This dish is great for people's health. The two vegetable ingredients are popular seasonal foods. Water chestnuts help clear body heat and produce saliva, and winter bamboo shoots reduce phelgm and swelling. Pork nourishes yin and clears body heat as well. So this dish is a good try in the height of winter. In addition, puchiin in water chestnuts prevents bacteria and diminishes inflammation. Winter bamboo shoots have rich cellulose and reduce cholesterol and blood fat.

腊味八宝粥

Laba Congee

食材配方

珍珠米300克　黑糯米50克　糯米100克　小红枣30个　花生80克　赤小豆80克　莲子40粒　百合30克　桂圆肉30克　冰糖150克

制作工艺

① 将莲子、小红枣、百合用温热水浸泡至软；桂圆肉改刀成小块；花生浸泡后去掉红衣。

② 锅置火上，加入清水、珍珠米、黑糯米、糯米、赤小豆，先用大火烧沸后改用小火熬煮，至粥汤起黏后加入花生、小红枣、莲子、桂圆肉、冰糖，先用中火烧沸，再改成小火熬至粥汤浓稠后起锅装碗即成。

风味特色

口感软糯润滑，味道甜香，老少皆宜。

创意设计

小寒常常与腊八节相邻。俗话说“腊七腊八冻掉下巴”，在这寒冷之季，人们习惯将五谷杂粮融合在一起做成腊八粥食用。在养生方面，此粥粗细搭配、蛋白质互补，食疗效果颇佳。其中，红枣补气养血、养心安神，花生健脾养胃、润肺化痰，莲子健脾益肾，百合养阴润肺、清心安神，与米、豆共煮，可调补气血、养阴生津。

Ingredients

300g pearl rice; 50g black glutinous rice; 100g glutinous rice; 30g small red dates; 80g peanuts; 80g red beans; 40 lotus seeds; 30g lily bulbs; 30g longan pulp; 150g rock sugar

Preparation

1. Soak the lotus seeds, red dates and lily bulbs in warm water till soft. Dice the longan pulp. Soak the peanuts and remove the red peels.
2. Add water, pearl rice, black glutinous rice, glutinous rice, and red beans to a pot, bring to a boil over high heat, turn down the heat and simmer till the soup turns glutinous. Add peanuts, red dates, lotus seeds, longan pulp and rock sugar, bring to a boil over a medium flame, and simmer over a low flame till the soup is thick. Transfer to a serving bowl.

Features

soft, glutinous and smooth rice; sweet and fresh tastes; popular with the old and young.

Notes

Lesser Cold, the twenty-thrid solar term in the Chinese lunar calendar, is near Laba (the eighth day of the last lunar month) Festival. As a Chinese saying goes, “On the seventh or eighth day of the twelfth month of the lunar year, it is so freezing cold that your teeth chatter”. In cold winter, people have the custom of making Laba Congee with various cereals. The dish is great for people’s health. Various cereals provide rich protein. Particularly, red dates replenish qi, nourish the blood and ease the nerves. Peanuts tonify the spleen, the stomach and the lung, and reduce phelgm. Lotus seeds tonify the spleen and the kidney. Lily bulbs nourish yin and the lung, and clear away the heart heat. All those ingredients cooked with rice and beans help tonify qi and nourish the blood, nourish yin and produce saliva.

GREATER COLD

大寒
枝枯桠疏
孤鹊独飞
大寒吟
宋·邵雍
旧雪未及消，
新雪又拥户。
阶前冻银床，
檐头冰钟乳。
清日无光辉，
烈风正号怒。
人口各有舌，
言语不能吐。

大寒物候 | PHENOLOGY IN GREATER COLD

大寒是二十四节气中的最后一个节气。当太阳到达黄经300° 时为大寒，时间为每年的1月19日～21日。明朝王象晋《群芳谱》言："大寒，寒威更甚。"

Greater Cold is the last of the 24 Solar Terms. It begins when the sun reaches the celestial longitude of 300°. It usually begins from 19th to 21st January every year. In his book *An Encyclopedia of Plants*, Wang Xiangjin in Ming Dynasty said, "Greater Cold, it is extremely cold".

我国古代将大寒后的十五天分为三候：一候鸡乳；二候征鸟厉疾；三候水泽腹坚。意思是说，大寒节气后五日，母鸡开始孵小鸡；再过五日，鹰隼之类的征鸟，正处于捕食能力极强的状态，到处寻找食物，以补充能量抵御严寒；又经五日，江河里的冰一直冻到江河中央，并且冻得最坚硬、最厚实。大寒相应的花信为：一候瑞香；二候兰花；三候山矾。

Ancient China has divided the 15 days after Greater Cold into 3 pentads. In the first pentad, hen has hatched chicken. In the second pentad, raptors like eagle and falcon, whose abilities of predation are at the peak, are looking for foods everywhere to replenish energy and overcome the freezing temperature. In the third pentad, the center of river have been frozen. Ice there is the hardest and thickest. The news of flowers blooming in Greater Cold are as follows: daphne for the first pentad, orchid for the second pentad and symplocos caudata for the third pentad.

食材生产 | PRODUCING FOOD INGREDIENTS

大寒时节是一年中最寒冷的节气。此时，北方地区早已进入农闲时间，南方的食材生产也相对轻松一些。在四川，一些浅山丘陵地带的稻田已蓄水越冬，以备来年春季耕种，平原地区的果蔬生长需要加强管理。此时正值隆冬，也是畜禽疫病多发季节，人们还要积极做好畜禽的保暖防冻与疾病防控工作。四川地区在此时出产的蔬菜主要有菠菜、韭黄、茼蒿、青油菜、红油菜薹以及草莓和名为"春见（耙耙柑）"的柑橘类水果。

Greater Cold is the coldest solar term in a year. The northern area is in winter slack season, and the food ingredients producing in the southern area is comparatively relaxed in this period. In Sichuan, people has stored water in rice fields for the coming spring ploughing and spring sowing in hilly area. They have strengthened the management of the growth of fruits and vegetables in plain. It is midwinter, the disease season for livestock and poultry. People should protect livestock and poultry from coldness and disease. Seasonable vegetables in Sichuan region are spinach, chive, crowndaisy chrysanthemum, green rape, bolt of red rape, strawberry and local citrus named "Chun Jian" or "PaPa Gan"

饮食养生 | DIETARY REGIMEN

大寒是极寒之时，但也是冬季即将结束之时，"寒极春来"，蕴含大地回春的迹象。它是一个季节交接的过渡节气，大寒过后不久就要迎来立春。此时，人体一方面需要有充足养分抵御风寒，另一方面又要补气养血，助阳气生发，为春季积蓄能量。在饮食上应遵循养肾、养藏、养阴

《年节习俗考全图·祭灶神》

的总原则，还需要特别重视养脾脏，并适当调养肝血，宜热食，少食寒凉、生冷食物，以免损害脾胃阳气，可适当食用味浓厚、有一定热量和脂肪的食物，还应多食用黄绿色蔬菜，如胡萝卜、油菜、菠菜等。此外，可多吃山药、山楂、柚子等具有健脾消滞功效的食物，也可多吃小米粥、雪梨柚子茶、雪梨双耳（银耳、木耳）羹等进行调理。

It is extremely cold in Greater Cold and it is the end of Winter. "Spring will come after the peak of coldness". It indicates the return of spring. Greater Cold is a transition solar term for the change of seasons. The Beginning of Spring just follows Greater Cold. At this moment, on the one hand, human body needs abundant nutrition to overcome the freezing temperature; on the other hand, people should tonify body energy, help yang qi to grow and reserve energy for spring. As for the diet, we should follow the primary principle of emphasizing on spleen, tonifying liver blood, taking more warm foods and less raw cold foods. We can eat some thick foods with proper calorie and fat. We can also take more green and yellow vegetables like carrot, rape, spinach, etc. Furthermore, people should take more Chinese yam, haw, pomelo which can strengthen spleen and remove food retention. People can eat more millet porridge, pomelo tea with snow pear, snow pear soup with tremella and agaric to recuperate.

大寒美食 | FINE FOODS IN GREATER COLD

民谚说“大寒过年”。大寒节气的时间，常常与岁末时间相重合，接近春节，所以，这个节气已充满了浓郁的年味，同时伴有丰富的民俗活动，其核心就是迎接新年的到来。农历腊月二十三，民间称为“过小年”，主要的民俗活动有扫尘、洁物、贴窗花、祭灶等。民间认为，“灶神”是掌管一家人一年祸福的神灵，腊月二十三灶神要升天向玉皇大帝汇报一家功过，因此，每逢此日，人们就用纸马、饴糖、米粑粑等送“灶神”上天，称为“送灶”，到除夕日又贴上新的灶神像，谓之“迎灶”。祭灶在于祈求新年合家平安如意。在四川，祭灶的日子一般是在腊月二十三号或二十四号。据《温江县志》载：“十二月二十四日，旧传灶神上天奏事，先于二十三日夜，各具香花、酒果、灯烛、饧糖等致祭。”

As the proverb goes, "celebrate the Spring Festival in Greater Cold". Normally, the date of Greater Cold is at the end of a year, close to Spring Festival. Hence, Greater Cold is full of atmosphere of Spring Festival. There are lots of folk activities, and the core is welcoming the arrival of the new year. On 23rd December in lunar year, people will celebrate Minor Spring Festival (usually a week before the lunar New Year). Main folk activities on that day are sweeping dust, cleaning INGREDIENTS, sticking window paper-cut, worshiping Kitchen God, etc. Chinese people believe that the Kitchen God is in charge of the family's fortune in a whole year, and he will report the family's deeds to Jade Emperor in Heaven on 23rd December. Therefore, people will offer paper horse, maltose and rice cake to Kitchen God on that day every year, which is called "Song Zao" in Chinese. On New Year's Eve, Chinese people will post new image of Kitchen God, which is called "Ying Zao" in Chinese. The purpose of worshiping Kitchen God is to pray for the peace and good luck of the whole family in the coming new year. In Sichuan, the day of worshiping Kitchen God is 23rd or 24th December in lunar year. *Wenjiang County Local Chronicles* recorded, "On 24th December, Kitchen God will go to heaven to give report to Jade Emperor. People will offer fragrant flower, liquor, fruit, candle, lamp, maltose to Kitchen God in the evening of 23rd December."

由于大寒节气多适逢岁末时节，除了食用具有滋补作用的美食外，很多家庭都陆续开始准备和食用丰富的过年应节食物，南方地区喜食各种羹汤及糯米饭以驱寒保暖、滋补身体。在四川

等地，还有“打牙祭”的习俗。“做牙”又称“牙祭”，是指传统民俗中每月两次祭祀土地神的活动。民间认为，土地神能够保佑商家生意兴隆，高朋满座，客似云来。故在农历每月的初二、十六日“做牙”，主要用鸡肉、猪肉、鱼肉等肉类食品祭拜土地神。一般是以农历二月二日土地神的生日为起点开始“做牙”，也称为“头牙”，十二月十六日这一天的“做牙”则称为“尾牙”，更是要进行“尾牙祭”。这一天，工商之家要清结账目，召集员工吃一顿丰盛的团年宴，然后放假，各自回家过年。到了现代，大寒时节的“尾牙祭”已演变为众多企业在岁末年初举办的庆祝年会，通过聚餐来犒劳员工、激励士气。

Because Greater Cold is at the end of a year, lots of families have started to prepare and eat abundant seasonal foods for the new year, in addition to taking tonic foods. In Sichuan, there is a custom of having a rare sumptuous meal, which is a activity of worshiping God of Land twice a month. Common people believe that God of Land can bless the merchants, can make the business flourish. Hence, on the 2nd day and 16th day of a month, people will offer chicken, pork, fish and other meat to God of Land, which we call “Zuo Ya” in Chinese. Normally, the first “Zuo Ya” in a year is 2nd February in lunar year, which is the birthday of God of Land. That is also called “Tou Ya”. Zuo Ya on 16th December in lunar year is called “Wei Ya” in Chinese and there will be a ceremony. On that day, businessman will provide an abundant banquet to his staffs. After the banquet, employees will go back home to welcome the new year. In modern times, that ceremony has become the annual conference for many enterprises. Entrepreneurs will reward and inspire his staffs by hosting a dinner party.

年年有鱼

Good Luck Fish

食材配方

翘壳鱼一尾（约1500克） 泡豇豆100克 葱花20克 泡姜20克 郫县豆瓣15克 泡辣椒末20克 食盐3克 味精2克 白糖20克 醋10毫升 酱油2毫升 姜片10克 葱节15克 姜米20克 蒜米20克 料酒30毫升 藿香碎10克 鲜汤300毫升 水淀粉35克 食用油100毫升

制作工艺

① 翘壳鱼宰杀后治净，从鱼背两边划开，加入料酒、食盐、姜片、葱节码味15分钟，入笼蒸12分钟至鱼肉成熟后取出装盘。

② 泡豇豆切成颗粒；泡姜切成米粒。

③ 锅置火上，入油烧至120℃时，下郫县豆瓣、泡辣椒末炒香出色，再加入姜米、蒜米、泡豇豆、泡姜米炒香，然后入鲜汤、白糖、酱油、醋、料酒、味精烧沸，用水淀粉勾芡，待汁浓吐油时加入藿香碎、葱花推匀，起锅浇淋在鱼身上即成。

风味特色

成菜色泽红亮，鱼肉质地细嫩，味道咸鲜微辣，略带甜酸。

创意设计

“鱼”与“余”谐音，用鱼烹制成菜，能寓意“年年有余”的美好愿望，因此，鱼便成了四川家家户户年夜饭中必不可少的主角。大寒是二十四节气中的最后一个节气，此时吃鱼，有祈盼来年生活“幸福美满，如鱼得水”等意。在养生方面，翘壳鱼开胃消食、健脾行水，符合年关寒冷时节的人体需要；具有四川特色的泡辣椒、泡菜可增开胃生津之功；藿香芳香化湿、行气消滞，更是烹制鱼肉的好搭档。

Ingredients

1 qiaoke fish (about 1,500g); 100g pickled yardlong beans; 20g spring onions, finely chopped; 20g pickled ginger; 15g Pixian chili bean paste; 20g pickled chilies, finely chopped; 3g salt; 2g MSG; 20g sugar; 10ml vinegar; 2ml soy sauce; 10g ginger, sliced; 15g spring onions, segmented; 20g ginger, finely chopped; 20g garlic, finely chopped; 30ml Shaoxing cooking wine; 20g huoxiang herb (agastache rugosus), finely chopped; 300ml stock; 35g average batter; 100ml cooking oil

Preparation

1. Slaughter and rinse the fish. Make two cuts into both sides of the fish back, and marinate with Shaoxing cooking wine, salt, sliced ginger and segmented spring onions for fifteen minutes. Steam the marinated fish in a steamer for twelve minutes until cooked through, and transfer to a serving dish.
2. Cut the pickled yardlong beans and pickled ginger into rice-size grains.
3. Heat oil in a wok to 120℃, and stir fry Pixian chili bean paste and pickled chilies till fragrant. Blend in the chopped ginger, garlic, yardlong beans, pickled ginger, and continue to stir till aromatic. Add the stock, sugar, soy sauce, vinegar, cooking wine and MSG, bring to a boil, and pour in the batter. Braise till the sauce is thick and lustrous, add huoxiang herb and chopped spring onions, stir well and pour over the fish.

Features

reddish brown color; tender fish; savory, spicy, sweet and sour tastes.

Notes

In Chinese, "fish" and "abundance" have the same pronunciations, so fish is a must for the New Year Eve feast in Sichuan. The Greater Cold is the last of the twenty four solar terms, having fish represents people's wish for abundance and happiness in the coming year. In terms of food therapy, qiaoke fish helps with stomach digestion and spleen function, and Sichuan pickles arouse the appetite. Huoxiang herb removes inner humidity and promotes qi circulation, a perfect match for fish.

腊味拼盘

Assorted Cured Meat

食材配方

老腊肉150克　腊猪舌150克　腊香肠150克　腊猪耳150克　花生仁250克
四川老卤水750毫升

制作工艺

① 将花生仁放入卤水锅中卤煮至软熟后捞出，放入攒盒中垫底。
② 老腊肉、腊猪舌、腊香肠、腊猪耳分别焯水，放入锅中煮熟（或蒸熟）后捞出，晾凉后切片，再将其分门别类摆放在花仁上即可。

风味特色

成菜色泽棕红，味道咸鲜，腊味浓香。

创意设计

大寒之时已近年关，腊味对于四川人而言，就意味着浓浓的年味，也融入了川人的丝丝乡愁。大寒是二十四节气的收关节气，在此时节食用腊味，除了可品味其独特的烟熏香美味道外，还有盼望团圆之意。就养生而言，此腊味拼盘有荤有素，不仅蛋白质含量丰富，花生仁所含的不饱和脂肪酸还能降低胆固醇及血脂。此外，烟熏所产生的特殊风味还能有效杀菌、防腐。

Ingredients

150g cured pork; 150g cured pork tongue; 150g cured sausages; 150g cured pork ear; 250g peanuts; 750ml Sichuan-style spiced broth

Preparation

1. Boil the peanuts in the broth till soft and cooked through, remove and transfer to the base of a cuanhe (a divided container).
2. Blanch the pork, pork tongue, sausages and pork ear, steam or boil till cooked through, remove and leave to cool. Slice them and place on top of the peanuts.

Features

appealing brown color; savory taste with peculiar aroma from curing

Notes

The Greater Cold, which is the last of the twenty-four solar terms, symbolizes that the new year is around the corner. Cured meat is always associated with new year celebrations in Sichuan, so it appeals to Sichuan people not only because it has a pleasant smoky taste, but because it reflects their wishes for family reunion. In terms of nutrition, the dish is a balanced combination of both meat and non-meat ingredients with high protein contents. Besides, the unsaturated fatty acid in peanuts lowers cholesterol and blood fat. The process of smoking helps with bacteria killing and food preservation.

饴香慈菇饼

Maltose-Flavor Water Chestnut Pies

食材配方

鲜慈菇250克　熟面粉25克　蜜枣120克　熟桃仁碎30克　猪油10克
饴糖50克　食用油1000毫升（实耗30毫升）

制作工艺

① 鲜慈菇去皮，入开水锅中煮熟后捞出剁成小颗粒，加入熟面粉拌匀制成慈姑面团。

② 蜜枣上笼蒸软，取出后压成泥，加入熟核桃碎、饴糖和猪油擦匀，制成枣泥馅，再分成每个重15克的馅心。

③ 将慈菇面团揪成每个重25克的剂子，包入馅心，收口后按成圆饼成饼坯。

④ 锅置火上，入食用油烧至150℃时，下饼坯炸至饼色微黄时起锅装盘，配上饴糖味碟即成。

风味特色

饼色美观，外酥内软，慈菇脆嫩，味道甘甜，饴香浓郁。

创意设计

四川方言中的慈菇、茨菇都是指的荸荠。此点以鲜慈菇为主料，包入带有饴糖香甜味的核桃枣泥馅心，风味独特。在养生方面，慈菇养阴，核桃温阳，面粉、蜜枣补气养血，符合此时节既需御寒又宜润燥的需要。此外，慈菇所含的荸荠英等成分，对抗菌、抗氧化、抗癌等也有一定的作用。

Ingredients

250g water chestnuts; 25g cooked wheat flour; 120g honeyed red dates; 30g cooked walnuts, crushed; 10g lard; 50g maltose; 1,000ml salad oil (30ml to be consumed)

Preparation

1. Peel the water chestnuts, add to boiling water and boil till cooked through. Remove, chop into small grains, and mix well with the wheat flour to make dough.
2. Steam the red dates in a steamer till fully cooked, remove, press and mash into a paste. Add the walnuts, maltose and lard, and blend well to make the filling. Divide into pieces about 15g each.
3. Cut the dough into 25g pieces, wrap up the filling, seal and press into round shapes.
4. Heat the cooking oil in a wok to 150℃, fry the round pieces of dough till golden, and transfer to a serving dish. Serve with maltose dipping sauce.

Features

pleasant golden color; soft pies with crispy skin and crunchy texture; a peculiar maltose flavor

Notes

This dish uses water chestnuts as the main ingredient for the wrapper and maltose, walnut and date as the fillings. In terms of food therapy, water chestnuts benefit yin while walnuts nourish yang. Wheat flour and honeyed red dates promote qi circulation and nourish the blood. The pies help to defend the cold and relieve the dryness of the season. Besides, water chestnuts contain puchiin which has antibacterial, anti-oxidation and cancer-prevention effects.

中华二十四节气菜（川菜卷）

CHINESE 24 SOLAR TERM DISHES（SICHUAN CUISINE）